The ESSENTIAL REAL ESTATE DICTIONARY

es·sen·tial. ADJ. Of the utmost importance.

Lisa Holton

SPHINX® PUBLISHING
AN IMPRINT OF SOURCEBOOKS, INC.®
NAPERVILLE, ILLINOIS
www.SphinxLegal.com

Copyright © 2010 by Lisa Holton
Cover and internal design © 2010 by Sourcebooks, Inc.®

All rights reserved. No part of this book may be reproduced in any form or by any electronic or mechanical means including information storage and retrieval systems—except in the case of brief quotations embodied in critical articles or reviews—without permission in writing from its publisher, Sourcebooks, Inc.® All brand names and product names used in this book are trademarks, registered trademarks, or trade names of their respective holders. Sourcebooks and the colophon are registered trademarks of Sourcebooks, Inc.®

First Edition: 2010

Published by: **Sphinx® Publishing, An imprint of Sourcebooks, Inc.®**

Naperville Office
P.O. Box 4410
Naperville, Illinois 60567-4410
(630) 961-3900
Fax: (620) 961-2168
www.sourcebooks.com
www.SphinxLegal.com

This publication is designed to provide accurate and authoritative information in regard to the subject matter covered. It is sold with the understanding that the publisher is not engaged in rendering legal, accounting, or other professional service. If legal advice or other expert assistance is required, the services of a competent professional person should be sought.
From a Declaration of Principles Jointly Adopted by a Committee of the American Bar Association and a Committee of Publishers and Associations

This product is not a substitute for legal advice.
Disclaimer required by Texas statutes.

Library of Congress Cataloging-in-Publication Data

Holton, Lisa.
  The essential real estate dictionary / by Lisa Holton. -- 1st ed.
    p. cm.
  A rev. and updated ed. of: The essential dictionary of real estate. c2004.
  (pbk. : alk. paper) 1. Real property--Dictionaries. 2. Building--Dictionaries. I. Holton, Lisa. Essential dictionary of real estate. II. Title. III. Title: Real estate dictionary.
  HD1365.H64 2008
  333.3303--dc22

                                    2008040035

Printed and bound in the United States of America.
RRD 10 9 8 7 6 5 4 3 2

CONTENTS

Definitions 1
Abbreviations 403
Websites 409

DEFINITIONS

1031 exchange. See *tax-free exchange*.

72-hour clause. N. When a buyer still has a house to sell before he or she can purchase another home, most sellers insist on a 72-hour clause in case a better offer comes in. If such an offer comes in, the seller then goes back to the buyer and gives him or her the option to remove the contingency and make a deal or lose the house.

AAA tenant/borrower. N. A tenant or borrower with an exemplary credit rating.

abandonment. N. The voluntary surrendering of property rights to someone else without transferring title.

abatement. N. For real estate purposes, a reduction in tax, rent, or utility rates, typically offered as an incentive by landlords or local governments.

abeyance. N. Lapse in succession during which title to a piece of property is not clearly established.

above building standards. N. Upgraded features of a project beyond basic project requirements.

absentee owner. N. A landlord who lives elsewhere.

absolute priority rule. N. The idea that creditors' claims take precedence over shareholders' claims in the event of a liquidation or reorganization.

absolute title. N. A clear title without any liens or judgments.

absorbed. V. When the cost is treated as an expense, rather than passed on to customers. N. A business that is merged into another company due to an acquisition.

absorption rate. N. How a developer estimates the expected annual sales or occupancy of new homes or commercial property. For example, if there are one hundred new homes for sale in a given month and ten sell while another ten are built, that month's absorption rate is 10%.

abstract of judgment. N. A court judgment filed with the county that creates a lien against a piece of property.

abstract of title search. N. A historical review of public records to determine whether liens or defects of title exist on a piece of property to determine any complications that could interfere with clear ownership transfer. Done prior to closing of title on a sale.

abutting. V. When property adjoins or borders another property.

A/C. ABBRV. Air conditioner or air conditioning.

accelerated amortization. N. By paying additional principal on a mortgage or other loan, the borrower can shorten the effective term of the loan. As an example, a $100,000, thirty-year mortgage at 8% interest has a monthly payment of $733.76. By paying an additional $100 a month—thereby accelerating the amortization of the loan—a borrower can reduce the thirty-year term of the loan to twenty years. See also *additional principal payment*.

accelerated cost recovery system. N. A depreciation method used for most property placed into service from 1981 to 1986. This method allowed assets to be depreciated at a faster rate than had been allowed previously. The modified accelerated cost recovery system (ACRS) replaced the ACRS for assets placed into service after 1986. ABBRV. *ACRS*.

accelerated depreciation. N. A bookkeeping method that allows a taxpayer to depreciate property more quickly in the early years of ownership.

acceleration clause. N. A loan provision that allows the lender to declare the entire loaned amount due immediately if the borrower pays late or breaks other covenants.

accession. N. The right of an owner to have the advantages of property ownership, which include air rights, mineral rights, riparian rights, and man-made improvements.

accessory apartment. N. A separate apartment created within a single-family home and occupied by a family member or a renter. It is advisable to check with the local zoning board to see if these units are legal before building.

access right. N. An owner's right to get to and from his or her property.

accommodating party. N. The intermediary or facilitator in a 1031 or like-kind exchange. This individual or entity holds the property while the exchange is being completed.

A/C condenser. N. The outside fan of an air conditioning system. It removes heat from Freon gas and transforms the gas back into a liquid and pumps it back into the coil in the furnace.

account. N. A record of financial transactions for an asset or individual, such as at a bank, brokerage, credit card company, or retail store. More generally, an arrangement between a buyer and a seller in which payments are to be made in the future.

accountant. N. One who is skilled in the practice of accounting or who is in charge of public or private accounts.

accountant's letter. See *accountant's opinion*.

accountant's opinion. N. A letter preceding a financial report, written and signed by an independent accountant, that describes the scope of the statement and presents an opinion on the quality of the data presented. Synonymous with *accountant's letter*.

accounting. N. The systematic recording, reporting, and analysis of financial transactions of a business.

accounting equation. N. The fundamental balance sheet equation: assets – liabilities = net worth.

accounting method. N. The method under which income and expenses are determined for tax purposes. Major accounting methods are the cash method and the accrual method.

accounting period. N. The twelve-month period that a taxpayer uses to determine federal income tax liability. Unless a taxpayer makes a specific choice to the contrary, his or her accounting period is the calendar year.

account reconciliation. N. The act of confirming that the balance in one's checkbook matches the corresponding bank statement.

accounts payable. N. Money that a company owes to vendors for products and services purchased on credit.

accounts receivable. N. Money owed to a company by a customer for products and services provided on credit. Treated as a current asset on a balance sheet.

account statement. N. A record of transactions and their effect on account balances over a specified period of time, for a given account.

Accredited Personal Financial Planning Specialist. N. A certified public accountant who has passed a rigorous financial planning examination.

accrual. N. The recognition of revenue when earned or expenses when incurred regardless of when cash is received or disbursed.

accrual accounting. N. The most commonly used accounting method, which reports income when earned and expenses when incurred, as opposed to cash basis accounting, which reports income when received and expenses when paid.

accrued expense. N. An expense that is incurred, but not yet paid for, during a given accounting period.

accumulated depreciation. N. The amount of depreciation expense that has been claimed to date.

A/C disconnect. N. The main electrical on-off switch near the A/C condenser.

acid-test ratio. See *quick assets ratio*.

acknowledgment. N. Declaration, in writing, that a person has acted voluntarily and is usually verified by an authorized official.

acquest. V. To acquire property through purchase.

acquiescence. N. Accepting or complying without objection, thus implying the waiver of the right to legal action.

acquisition. The securing of ownership or controlling interest in a property or other object through either purchase or a merger.

acquisition cost. The total cost of property and the fees incurred in buying it. If the selling price of a property is $100,000 and the fees to buy it totaled $5,700, the total acquisition cost of the property is $105,700.

acquisition debt. N. Debt incurred to acquire, construct, or improve the taxpayer's principal or secondary residence.

acre. N. A two-dimensional land measurement equaling 43,560 square feet.

ACRS. ABBRV. Accelerated Cost Recovery System.

action. N. A procedure brought before a court, in the form of a complaint, to demand a legal right, which in real estate would be to repossess or regain certain properties.

action in personam. N. Judicial proceeding against a person rather than against the property of that person. It would seek to have that person uphold the terms of a contract, make up for a loss, or provide a service. In common law, it seeks the payment for a debt or damages incurred.

action in rem. N. (*Latin*) Against the thing; judicial proceedings against property. While in legal theory an action in rem occurs only against property, in actuality it consists of a legal action between parties for the purpose of attaching or disposing of property owned by them.

active asset. N. An asset used in the daily operations of a business.

active income. N. Income for which the taxpayer performs services. Examples are wages, salaries, tips, bonuses, and business and partnership income in which the taxpayer materially participates in the business or partnership. See also *income, passive income, portfolio income*.

active participation. N. Involvement in real estate ownership and management on a continuing basis as opposed to engaging a property manager or some other intermediary to handle all the work and details of owning property. Tax laws and lenders provide greater benefits when the owner actively participates in real estate property and rentals. See also *sweat equity*.

act of God. N. Action occurring without the intervention of man, which could include but not be limited to hurricanes, earthquakes, floods, and lightning.

act of God delay. See *force majeure*.

actual eviction. N. The act of removing, dispossessing, or expelling an individual from premises by force or by law.

actual notice. N. Confirmation that the purchaser has been shown all the relevant details of a transaction. These details include all information in the public record relating to the title, e.g., liens, surveys, access, air rights, mineral rights, water rights, etc. Actual notice is either express or implied. Express notice means that the purchaser was shown all the relevant details. Implied notice assumes that the purchaser's knowledge was sufficient to institute investigation and inquiry.

ADA. ABBRV. Americans with Disabilities Act.

addendum. N. Contractual changes or additions; a supplement or addition to a transaction agreement of any kind.

additional principal payment. N. Payment made in excess of the required monthly payment, therefore reducing the principal and shortening the term of the loan. See also *accelerated amortization*.

add-on-interest. N. The amount of interest paid on the principal of a loan for the duration of that loan.

adhesion contract. N. A legally enforceable contract that is offered on a take-it-or-leave-it basis, leaving no opportunity for the purchaser to negotiate and must be accepted as is. Depending on the item, it could put the purchaser at a distinct disadvantage.

ad hoc. ADV. (*Latin*) For this; when something is used for the purpose at hand and not considered for a wider application.

ad infinitum. ADJ. (*Latin*) Without end. ADV. ***ad infinitum***.

adjacent property. N. Property that is close to but not touching another property.

adjoining owner. N. A property owner with property touching a common property and who has a legal right to notification when a zoning variance or change to the common property is being formally considered.

adjoining property. N. Contiguous property sharing a common border.

adjournment of closing. N. Postponement of closing of title until another day or place. This can occur for various reasons, e.g., the absence of one party or incomplete documents.

adjudication. N. The formal court decision taken in an action, which has been brought reaching a final determination for all parties involved.

adjudication order. See *judgment*.

adjustable rate mortgage. N. A loan with a periodically adjustable interest rate, reflecting the changes in a specific financial index. Adjustable rate mortgages are typically offered with one, three, or five-year periods before the rate can change. Before a borrower signs, it is good to know how the interest rate will change when it does. An interest rate cap limits the amount by which the interest rate can change when the lock period ends; look for this feature in considering an adjustable rate mortgage loan. The following chart illustrates the popularity of adjustable rate mortgages. ABBRV. *ARM*.

1- to 4-Family Mortgage Originations, 1990-2001			
Year	Total Volume (Millions)	Refi Share (Percent)	ARM Share (Percent)
1990	458,404	15	28
1991	562,074	31	23
1992	893,681	47	20
1993	1,019,861	52	20
(cont.)			

Year	Total Volume (Millions)	Refi Share (Percent)	ARM Share (Percent)
1994	768,748	24	39
1995	639,436	21	33
1996	785,233	29	27
1997	833,650	29	22
1998	1,507,000	50	12
1999	1,285,000	34	22
2000	1,024,000	19	25
2001	2,030,000	57	12

Note: ARM Share is percent of total volume of conventional purchase loans.

Source: HUD Survey of Mortgage Lending Activity, Mortgage Bankers Association of America, Federal Housing Finance Board

adjusted balance method. N. A technique for calculating finance charges (such as in a bank account, charge account, or credit card account) based on the remaining account balance after adjustments are made for payments and credits during the billing period. Interest charges are usually lower under this method than under other methods, such as average daily balance and previous balance methods.

adjusted basis. N. The base price of an asset or security that reflects any deductions taken on or improvements to the asset or security; used to compute the gain or loss when sold.

adjusted book value. N. The book value on a company's balance sheet after assets and liabilities are adjusted to market value. Property is typically part of this calculation.

adjusted cost basis. N. Tax purpose cost measuring method that allows cost to be increased by the cost of capital improvements or reduced due to depreciation. The cost of a home would be increased by the amount paid to install a permanent improvement such as air conditioning.

adjusted gross income. N. Federal tax term that refers to the difference between a taxpayer's gross income and adjustments to that income. These adjustments may include deductions for individual retirement accounts and Keogh pension plans, which may be invested in real estate. Adjusted gross income is the basis for determining the limitations or eligibility of other components, such as miscellaneous expenses (2% of adjusted gross income), in calculation of taxpayer's tax liability. ABBRV. *AGI.*

adjusted price. N. Price plus accrued interest.

adjusted sales price. N. Net sales price of a piece of property with the commissions and closing costs deducted from the actual sales price. For appraisal purposes, it is the price of a comparable piece of property with the differences between it and the property in question accounted for. For income tax purposes, it is the sales price of a property reduced by the costs of improvements necessary to bring it to the condition of the comparable property.

adjusted tax basis. N. The amount used to determine profit or loss from a sale or exchange of property. To determine an adjusted basis for an asset, start with the original cost. Add cost of improvements and assessments to the asset, and subtract deductions taken, such as depreciation and depletion. Example: A property was purchased for $300,000 and the owner made $100,000 worth of capital improvements. With an accumulated depreciation of $75,000, the adjusted cost basis for tax purposes would be $325,000 ($300,000 + $100,000 – $75,000). If the property were sold for $400,000, the taxable gain would be $75,000.

adjuster. N. The insurance company employee who settles the value of any damage done to insured property so that the insured owner may be compensated.

adjusting entry. N. A bookkeeping entry made at the end of an accounting period to correct the data for, and assign income and expenses to, a prior period.

adjustment. N. A deduction made to charge off a loss, as with a bad debt.

adjustment date. N. The date on which the interest rate changes for an adjustable rate mortgage.

adjustment interval. N. The time when changes in the interest rate or monthly payment are made on an adjustable rate mortgage loan.

adjustments. Items paid by the seller in advance and items yet to be paid for which the seller is responsible. The most common expense is property taxes, but others may have to be addressed. See also *buyer costs*.

adjustments in appraisal. N. Change in the value of one particular property based on the value of a comparable property. Adjustment in the appraisal of a comparable property will then affect the value of the original property.

administrator. N. A court-appointed individual who manages and distributes the estate of a person who has died without a will.

administrator's deed. N. The deed used by the court-appointed administrator of an estate to transfer property.

ADR. ABBRV. Asset depreciation range.

adult communities. N. Housing projects specifically designed to meet the needs of mainly older, retired persons with no small children.

ad valorem. ADJ. (*Latin*) According to value.

ad valorem tax. N. A tax based on the value of the item being taxed—governments and school districts raise most of their revenues from property taxes. Example: If the ad valorem tax rate were 2%, the tax would be $2 per $100 of assessed value.

advance. N. Increase in the price or market value of real estate or, most commonly, money given to someone before it is earned, as in payment for services or goods received prior to receipt of it.

adverse land use. N. The use of real estate contrary to the interests of the property owner and implying a lack of knowledge by the owner.

adverse possession. N. The acquisition of title to property through possession for a certain period of time set in state law, without knowledge by the owner. Adverse possession is a statute of limitations that prevents a legal owner from claiming title to the land when the owner has done nothing to evict an adverse occupant during the statutory period.

adverse use. N. Use and access of a certain property without the consent of the owner.

advocacy role. Representing the best interests of a client.

aerator. N. The round-screened, screw-on tip of a sink spout. It mixes water and air for a smooth flow.

aerial photos. N. Photographs of land areas and buildings taken by cameras mounted in airplanes or satellites. Developers, builders, civil engineers, geologists, geographers, and archaeologists use aerial photographs.

aesthetic value. N. The enhancement of a property's value by appearance or favored location.

affidavit. N. A legal document that proves that the signer understands various important parts of an agreement. A homebuyer will sign a variety of affidavits at closing, such as an affidavit of occupancy, which states that the signer will use the property as a principal residence. A buyer and seller may have to sign an affidavit stating that all the improvements to the property required in the sales contract were completed before closing. Lenders can answer questions about these documents before closing.

affidavit of title. N. Document in which the seller identifies him- or herself and swears to his or her marital status and that he or she is in possession of the property and certifies that, since the date of the examination of title, there have been no judgments, divorces, unrecorded deeds, unpaid repairs, bankruptcies, or defects in title that are known to him or her.

affirm. V. To confirm, ratify, verify, and accept a transaction, which can be canceled.

affirmation. N. An alternative to an oath, which can be used on people whose religious beliefs will not allow an oath.

affirmative lending. N. Based on the Community Reinvestment Act, affirmative lending is intended to increase the availability of loan dollars in poorly served neighborhoods.

affordability analysis. N. A detailed analysis of a borrower's ability to afford the purchase of a home. An affordability analysis takes into consideration income, liabilities, and available funds. It also considers the borrower's ability to pay for a particular type of mortgage and the closing costs associated with it.

affordable housing. N. The public and private sector movement to help low- and moderate-income people buy homes.

afforestation. N. Creating a forest cover on land area not previously covered with trees to preserve the ecology and increase the aesthetic value of the land.

A-frame. N. Outer shape of a structure that has steeply sloped roofs and is in the shape of an "A."

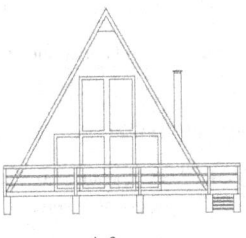

A-frame

after-completion costs. N. Expenditures incurred after the completion of a building.

after-tax cash flow. N. Cash flow (from income-producing property) that is reduced by income taxes resulting from the property's income.

after-tax equity yield. N. Net return rate, after deduction of interest costs and taxes, on an equity investment in real estate.

after-tax rate of return. N. Rate of return on investment once income taxes are deducted.

age-life depreciation. N. Depreciation method based on the projected useful life of the property, allowing for normal wear and tear.

agency. N. The relationship between two people or entities where one is a principal and the other is an agent representing the principal in activities with other parties, such as a real estate agent buying or selling for a client.

agency closing. N. The use of a title company or another firm to complete a loan.

agency disclosure. N. Law requiring real estate agents to disclose whether they represent the buyer or seller; this varies by state.

aggregate. N. A mixture of sand and stone and a major component of concrete.

AGI. ABBRV. Adjusted gross income.

agreed boundary. N. A compromise reached by two property owners to resolve a dispute over borders.

agreement of sale. N. A document detailing the terms and sale price of a transaction, along with a description of the property and any time limits. A contract obligates the buyer to buy and the seller to sell and is normally accompanied by a deposit from the buyer. Synonymous with *contract to purchase*.

agricultural property. N. Property zoned for use in farming, including the raising of crops and livestock.

agricultural use value. N. Determination of value of agricultural land by taking into consideration the amount of arable land (property not disrupted by trees or overgrowth or untillable terrain), its proximity to an adequate water supply, and the climatic location.

AI. ABBRV. Appraisal Institute.

AIA. ABBRV. American Institute of Architects.

air rights. N. Rights to use the open air space above a street, railroad line, or other property; sometimes used to allow construction of billboards or other signage. Air rights may also allow or prevent the construction of entire buildings, such as when a development is constructed over an existing highway.

air space. N. The area between the insulation and the interior of exterior wall coverings. Normally a one-inch air gap.

a la carte real estate service. N. Transactions rendered one at a time instead of a commission-based, full-service relationship.

alienation clause. N. A mortgage provision requiring that the balance of that loan be repaid in full if the property is sold or transferred.

allegation. N. A charge, claim, accusation, or statement of a party to an action against a respondent. The allegation establishes the circumstances to be proved in a formal action before the court.

all-inclusive trust deed. N. A mortgage (trust deed) that encompasses all existing mortgages and is subordinate to them.

allocation method. N. An approach used to allocate the price for two or more properties that is based on their fair market value rather than appraised value.

allodium. N. Piece of property that is owned outright with nothing due to another party.

allotment. N. Allocation of real estate made on an equitable basis to prospective buyers. See also *apportionment*.

allowable span. N. The maximum allowable distance between structural supports.

allowance. N. Amount of money offered by builders of new homes to be applied toward the cost of items subject to customer selection, such as lighting fixtures or carpeting.

allowance for depreciation. N. An accumulated expense that writes off the cost of a fixed asset over its expected useful life.

allowance for vacancy and income loss. N. The allowance made for the income lost from unoccupied premises when valuing income-producing property.

ALTA. ABBRV. American Land Title Association.

ALTA Title Policy. N. American Land Title Association policy that covers the lender's interest and provides more coverage than a regular owner's title policy. See also *American Land Title Association*.

alteration. N. Changes or modifications to a building or structure.

alternative documentation. N. A method of documenting a loan file that relies on information the borrower is likely to be able to provide instead of waiting on verification sent to third parties for confirmation of statements made in the application.

alternative minimum tax. N. Tax designed to prevent taxpayers from escaping a fair share of tax liability by use of certain tax breaks. A taxpayer is subject to this tax if he or she has certain minimum tax adjustments or tax preference items and his or her alternative minimum taxable income exceeds the exemption allowed for his or her filing status and income level. ABBRV. *AMT*.

alternative mortgage. N. A loan other than a standard fixed-rate mortgage. An alternative mortgage may vary in interest rate, maximum loan amount, or loan-to-value ratio.

amenity. N. A feature of real property that enhances its attractiveness and increases the occupant's or user's satisfaction even though that feature is not essential to the property's use. Natural amenities include a pleasant or desirable location near water, upgraded landscaping, or scenic views. Human-made amenities include swimming pools, tennis courts, community buildings, and other recreational facilities.

American Institute of Architects. N. A professional society of architects founded in 1857. Located in Washington, D.C., with 301 local groups in all fifty states and over fifty-four thousand members, it promotes excellence and professionalism in the field. ABBRV. *AIA*.

American Land Title Association. N. An organization located in Washington, D.C., with over two thousand, four hundred members in forty states that was founded in 1906 and fosters uniformity and quality in title abstract ad insurance policies. Its publications are the monthly *Capital Comment*, the bimonthly *Title News*, and the annual *Directory of Members*. ABBRV. *ALTA*.

American Motel Hotel Brokers Network. N. An association of real estate brokers who specialize in the sale and purchase of lodging institutions.

American Planning Association. N. An organization founded in 1978 to encourage the best techniques and decisions for the planned development of communities and regions. Located in Washington, D.C., it has twenty-six thousand members with forty-five regional groups in the United States. ABBRV. *APA*.

American Society for Testing Materials. N. A nonprofit organization of a variety of qualified professionals who decide the level of quality that must be used in building materials for a particular job. A standard is proposed and these professionals either approve it or suggest changes. There is a vote to decide on approval and then the decision is published, with which manufacturers are expected to comply. If they do not, they are not given an American Society for Testing Materials (ASTM) number and cannot be used for jobs that require ASTM-approved products. ABBRV. *ASTM*.

American Society of Appraisers. N. A society that is primarily concerned with the advancement of the appraisal profession including teaching, certifying, and testing. Located in Washington, D.C., it was founded in 1952 and has six thousand members. ABBRV. *ASA*.

American Society of Farm Managers and Rural Appraisers. N. An organization that focuses solely on the appraisal of farm and rural properties.

American Society of Home Inspectors. N. Professional association of independent home inspectors. These members must meet the group's education and performance requirements. ABBRV. *ASHI*.

American Society of Real Estate Counselors. N. Real estate professionals who provide counseling on real estate purchases and investment decisions through a negotiated fee rather than a commission. Founded in 1953 with 850 members, it is located in Chicago, Illinois. ABBRV. *ASREC*.

Americans with Disabilities Act. N. A law that makes it illegal to discriminate against a person with a disability in housing, public accommodations, transportation, employment, government services, and telecommunications. The act is part of a broad series of laws requiring commercial and residential builders to make construction accommodations for the disabled. ABBRV. *ADA*.

amortizable expenses. N. Certain capital expenses that can be deducted over a fixed period of time. They include start-up expenses, qualified forestation or reforestation costs, goodwill, going-concern value, covenants not to compete, franchises, trademarks, and trade names.

amortization. N. Regular payment of both principal and interest on a loan. Early in the loan, most of the payment is applied toward interest with increasing amounts paid toward principal as the loan moves toward maturity.

amortization schedule. N. A written schedule of all payments to be made on an amortized loan. It should show the date of each payment, the amount that will be applied to interest and principal each month, and the amount outstanding after that particular month's payment has been made.

amortization tables. N. Mathematical tables used to calculate a monthly payment on a loan.

amortization term. N. The amount of time required to amortize a mortgage loan. The amortization term is expressed as a number of months. For example, a thirty-year, fixed-rate mortgage has an amortization term of 360 months.

amortize. V. To gradually eliminate a financial obligation through periodic payments.

amortized mortgage. N. A mortgage where the interest and principal have been fully repaid by the mortgagee.

amount realized. N. The amount received by a taxpayer on the sale or exchange of property. The amount received is the sum of the cash and the fair market value of any property or services plus any of the seller's liabilities assumed by the purchaser. Determining the amount realized is the starting point for arriving at realized gain or loss.

anchor bolts. N. Bolts that secure a wooden sill plate to concrete, or a masonry floor or wall.

anchor tenant. N. A major tenant in a shopping center that helps the center attract shoppers who benefit the other tenants as well, e.g., a supermarket.

ancillary administrator. N. An out-of-state or out-of-jurisdiction administrator who is appointed to probate a decedent's property when there is no executor.

ancillary tenant. N. A tenant that occupies less space than an anchor tenant in a shopping center, e.g., a specialty store in a shopping mall.

annexation. N. A legal process used by a municipality to expand its territory to include part of an adjacent unincorporated area.

annual assessments. N. Valuation on property for purposes of taxation, sewer charges, etc.

annual cap. N. Amount the interest rate of an adjustable rate mortgage can be raised or lowered in any consecutive twelve-month period.

annual debt service. N. Total annual interest and principal loan payments required on a loan.

annualize. V. To make calculations for a period of less than a year as if the period were a whole year.

annual mortgage constant. N. The ratio of annual mortgage payments divided by the initial principal of the mortgage; only applies to loans involving constant payment. Annual mortgage constant = annual debt service ÷ mortgage principal.

annual mortgagor statement. N. Statement sent to borrower, on a yearly basis, detailing principal remaining on the loan and amounts paid toward interest and taxes.

annual percentage rate. N. The rate a borrower actually pays, including interest, points, and loan origination fees when expressed as a percentage rate per year. On an adjustable rate mortgage, assumption is made that the loan's index remains the same as its initial value. ABBRV. *APR*.

annual percentage yield. N. The real rate of interest on a loan; the coupon rate divided by the net proceeds of the loan. ABBRV. *APY*. See also *effective interest rate*.

annuity. N. Fixed sums paid, at regular intervals, to an investor.

annuity due. N. Annuity where the payments are made at the beginning of the period, either monthly, quarterly, or yearly.

annuity factor. N. A calculation that shows the value of a property's income stream.

anticipated holding period. N. The amount of time one expects to own property as an investment.

anticipatory breach. N. Communication—usually by letter—informing the other party that the terms of their original contract will not be fulfilled.

APA. ABBRV. American Planning Association.

apartment. N. Unit of one or more rooms within a multifamily complex of similar units.

Appeals Board. N. Government entity with the authority to overturn earlier government rules and policy on zoning and other property-related issues.

application fee. N. A fee paid to the lender at the time of application for a loan. It may include charges for a credit report, property appraisal, etc.

apportionment. N. Division or assignment based on a plan or proportion as in prorating property expenses such as insurance and taxes between the buyer and seller.

apportionment clause. N. A contractual provision requiring apportionment.

appraisal. N. The market value of a property or home as supplied by a third party, usually a licensed professional.

appraisal approach. N. A method used to estimate the market value of a property. The three basic methods of valuing property are the replacement value method, which is estimating the amount it would cost to replace the property without considering depreciation; the comparable sales approach, which is estimating cost by comparing similar properties with each other; and, the appraisal method, which is determining interest income that a property would return to the investor.

appraisal fee. N. The fee charged by a licensed professional to estimate the market value of a piece of real estate.

Appraisal Institute. N. An organization of professional valuers of real estate based in Chicago. ABBRV. *AI*.

appraisal of damage. N. An assessment of property loss, taking into account the quality, quantity, and age of a property; usually done by a professional appraiser or the insurance industry.

appraisal report. N. A detailed written report that shows the value of a property based on an area's recent comparable sales. It also includes a description of the property and structures, street address, zoning allowed, assessed valuation and taxes, best use for the property, and information about the appraiser.

appraised value. N. Professional opinion of the market value of a home or property.

appraiser. N. A licensed professional who is permitted to do appraisals and appear as an expert witness in a court of law regarding the evaluating process as well as to give testimony concerning real estate and market value.

appreciation. N. Increase in the value of real estate over a period of time.

approval. N. A confirmation of an amount able to be borrowed by an individual, based upon assessment of his or her ability to repay said loan. Alternatively, an authorization obtained from governmental authorities for a building project to proceed.

approximate compound yield. N. Measure of the annualized compound growth of a real estate investment.

appurtenance. N. An item that is outside the property itself but is considered a part of the property and adds to its enjoyment, such as a right of way.

appurtenant structure. N. Structure not belonging to a property but considered a part of it through the use of an easement of common interest.

APR. ABBRV. Annual percentage rate.

apron. N. A trim board installed beneath a windowsill.

APY. ABBRV. Annual percentage yield.

aqueduct. N. Large pipe made for bringing water from a distant source.

arbitration. N. Dispute-resolving method involving a third-party decision.

arbor. N. A latticework structure holding vines or flowers. Alternately, an arbor is a specially planted area filled with trees.

architect

architect. N. A licensed designer of homes, buildings, and other structures.

Architect's Punch List. N. List of design items needing to be corrected or resolved prior to finalization of a building design.

architectural fees. N. The fees charged by an architect for services rendered. Charges range from per square footage, hourly, or as a percentage of the projected budget.

architectural shingles. See *laminated shingles*.

area wall. N. A retaining wall around an area, located below grade.

areaway. N. A below grade, open space that allows light or access to a basement door or window.

ARM. ABBRV. Adjustable rate mortgage.

ARM Index. N. The index used to adjust the interest rates on adjustable rate mortgages. T-bills and prime rate are usually used for this index, which is not controlled by individual lenders.

array. ADJ. Tax assessor term describing a certain category of properties sold within a given amount of time.

arrears. N. Late or overdue payments that are in default.

arroyo. N. A dry ravine, found in arid areas and formed by water runoff. Not suitable for building as they are prone to flooding with significant rainfall.

artesian well. N. Drilled well with water rising through the opening due to naturally occurring water pressure and without using the pipe usually inserted into the outlet to control the water flow.

ASA. ABBRV. American Society of Appraisers.

asbestos. N. Fire-resistant material, formerly used for insulation and some home products, which has been found to be a health hazard and is no longer used. See also *contingency*.

asbestos cement. N. Fire-resistant cement made of a combination of asbestos fibers and Portland cement.

as-built drawing. N. A drawing made to show the actual dimensions and locations of installations.

ASHI. ABBRV. American Society of Home Inspectors.

as-is condition. N. Transfer of title to a property in an existing condition with no warranties or representations.

asking price. N. The initial price asked for by the seller for a property.

aspect ratio. N. Ratio between the height of an object and its width.

ASREC. ABBRV. American Society of Real Estate Counselors.

assess. V. To estimate value. Alternately, to levy a tax or fee on property.

assessed value. N. Value placed on a home by a government tax assessor in order to calculate a tax base.

assessment. N. Estimated value of a piece of property. Alternately, a levy placed on a property in addition to taxes.

assessment cycle. N. Period of time a municipality allows between valuations of property for tax purposes.

assessment rolls. N. List of taxable properties as compiled by the tax assessor.

assessor. N. Official who determines the assessed value of a property.

asset. N. (1) Item of value, e.g., cash, securities, investments, real estate, etc. (2) Any item of economic value owned by an individual or corporation, especially that which could be converted to cash. On a balance sheet, assets are equal to the sum of liabilities, common stock, preferred stock, and retained earnings.

asset coverage. N. The extent to which a company's net assets cover its debt obligations, preferred stock, or both; expressed in dollar terms or as a percentage.

asset depreciation range. N. Used to determine class lives for property and equipment, this is a range of depreciable lives allowed by the Internal Revenue Service for a particular asset. ABBRV. *ADR*.

asset/equity ratio. N. Total assets divided by shareholder equity.

assign. V. To transfer certain rights to another.

assignment. N. The transfer of a mortgage from one person to another.

assignment of mortgage. N. A written record of the transfer of a mortgage from the old owner to the new owner.

assignor. N. The person transferring rights and interests in a property.

assumable mortgage. N. A mortgage that can be taken over (assumed) by the buyer when a home is sold. In most cases, the lender must approve the assumption. An assumption fee is usually paid to the lender—usually by the purchaser of the property—after the assumption of an existing mortgage. A borrower should evaluate the terms and conditions of an assumable mortgage to

assumption. N. The transfer of the seller's existing mortgage to the buyer. See also *assumable mortgage*.

assumption clause. N. A provision in an assumable mortgage that allows a buyer to assume responsibility for the mortgage from the seller. The loan does not need to be paid in full by the original borrower upon sale or transfer of the property.

assumption fee. N. The fee paid to a lender (usually by the purchaser of real property) resulting from the assumption of an existing mortgage. The assumption fee is the amount paid to a lender resulting from a buyer taking over the payments on a seller's existing loan. The purchaser of the property usually pays this fee. Buyers pay an assumption fee when they take over the payments on a seller's existing loan.

ASTM. ABBRV. American Society for Testing Materials.

astragal. N. A molding that is attached to one of a pair of swinging double doors against which the other strikes.

at-risk rules. N. Tax laws that limit the amount of tax losses an investor—particularly a limited partner—can claim.

attached garage. N. A garage that is part of the same building as the dwelling.

attic. N. A floor consisting of open space at the top of a house just below the roof; often used for storage.

attic ventilators. N. Screened openings provided to ventilate an attic space.

attorney at law. N. An individual who has been admitted to the bar and allowed to practice law in his or her state and may perform all the services necessary to represent clients.

attorney of record. N. An attorney whose name appears in the permanent records of a case.

Attorney's Opinion of Title. N. A written statement by an attorney after examination of public records and abstracts of title that, in his or her judgment, the title to a particular property is clear.

attornment. N. Agreement by a tenant to accept a replacement landlord.

auction market. N. A system in which properties are sold to the highest bidder.

audit. N. (1) An examination and verification of a company's financial and accounting records and supporting documents by a professional, such as a certified public accountant. (2) An IRS examination of an individual or corporation's tax return to verify its accuracy.

audited financial statements. N. A company's financial statements that have been prepared and certified by a certified public accountant.

auditor. N. An individual qualified (at the state level) to conduct audits.

auditor's report. N. A section of an annual report containing an accountant's opinion about the accuracy of its financial statements.

audit trail. N. A step-by-step record by which financial data can be traced to its source.

automated underwriting. N. The process used to analyze how a borrower has handled credit obligations in the past and whether he or she has the ability to repay the mortgage for which he or she is applying. This includes a full assessment of income and assets as well.

automatic renewal clause. N. Provision in an agreement that allows renewal at the end of its initial term, such as a lease.

automatic sprinkler system. N. Fire protection system, required in some types of buildings, that is activated automatically by intense heat.

average accounting return. N. A measure of the return on an investment over a given period, equal to average book value over the duration of the investment.

average collection period. N. The average time period for which receivables are outstanding. Equal to accounts receivable divided by average daily sales.

average office occupancy. N. Average number of business days an office is in use.

average price. N. Price of a home, as figured by totaling the sales prices of the houses sold in an area, and dividing that number by the number of homes.

avigation easement. N. Easement over private property near an airport, which limits the height of structures or trees.

B

back charge. N. Billings for work performed by the homeowner that should have been performed by the contractor. Owners bill back charges to general contractors, and general contractors bill back charges to subcontractors. Examples of back charges include charges for cleanup work or to repair something damaged by another subcontractor, such as a tub chip or broken window.

back-end fees. N. Commission received by a syndicator when real estate is sold. They are typically paid after the investors receive their initial investment plus return.

back-end ratio. N. Lender calculations by which debt (principal, interest, property taxes, insurance, and other monthly bills) is compared with gross monthly income.

backfill. N. The replacement of excavated earth into a trench around or against a basement or crawl space foundation wall.

backing. N. (1) Wall backing is frame lumber installed between the wall studs to give additional support for drywall or an interior trim-related item, such as handrail brackets, cabinets, and towel bars. (2) Carpet backing is lumber that holds the carpet fabric in place.

back out. V. Framing contractors step in once mechanical subcontractors (heating, plumbing, electrical) finish their phase of work before the insulation stage to get the home ready for city inspection. Framing contractors generally clean up anything disturbed by others so a rough frame inspection can be passed.

back taxes. N. Unpaid property taxes.

back title letter. N. Letter given by the title company to an attorney to aid in his or her examination of a title.

back-to-back escrow. N. Arrangements necessary when a person plans the sale of one property and the purchase of another simultaneously.

backup contract. N. A second contract to complete a real estate deal if the first one falls through.

bad debt. N. Business accounts receivable that have been included in income in a prior year that are uncollectible, legally binding debts owed to the taxpayer that are totally worthless and uncollectible, and debts the taxpayer must pay that he or she guaranteed in connection with his or her business or for a profit may be deductible as bad debts. Synonymous with *charge off*.

bad debt reserve. N. An amount set aside as reserve for bad debts.

bad faith. N. Intent to deceive from the beginning of a deal or contract.

bad title. N. A situation where clear ownership is marred by unsettled claims and liens and may prevent an owner from selling.

balance. N. The amount of money in an account, equal to the net of credits and debits at that point in time for that account.

balanced budget. N. A budget for which current expenditures are equal to or less than income. The concept is often discussed in reference to the federal government.

balance sheet. N. Statement of financial condition, which lists assets, liabilities, and stockholder's equity. The net worth of an individual is assets minus liabilities. Net worth is important because it determines the creditworthiness of a borrower.

ballast. N. A transformer that steps up the voltage in a florescent lamp.

balloon-framed wall. N. Framed walls (generally over ten feet tall) that run the entire vertical length from the floor sill plate to the roof.

balloon mortgage. N. A mortgage that has level monthly payments that will amortize it over a stated term (often thirty years), but that provides for a lump sum payment to be due at the end of an earlier specified term. It is ideal for borrowers who want to sell or refinance a home within seven years and want a low monthly payment during that time. The interest rate paid on a balloon mortgage is usually lower than a comparable thirty-year, fixed-rate mortgage. Most balloon mortgages do have a refinance option, but it is wise to double check.

balloon payment. N. The final lump sum payment of a balloon loan.

balustrade. N. Railing on a porch or stairway that is held up by a set of posts. Also known as balusters.

bank letter. N. A letter provided by a bank to another party to document the availability of funds by an individual or business.

bankrupt. ADJ. Being financially insolvent.

bankruptcy. N. (1) A legal procedure by which an insolvent debtor can be relieved of repayment of certain obligations. Bankruptcies, depending on the type of filing, will remain on a credit rating for seven to ten years. This may become a problem in obtaining financing. (2) For tax purposes, a formal petition filed in a Bankruptcy Court under Chapter 7, 11, 12, or 13 of Title 11 of the U.S. Code.

barge. N. A horizontal beam rafter that supports shorter rafters.

barge board. N. A decorative board covering a rafter.

base. N. A trim board placed against the wall around the room next to the floor. Synonymous with *baseboard*.

baseboard. See *base*.

baseboard electric heat. N. Centrally controlled heating unit, installed along the floor.

base loan amount. N. The total amount used to figure loan payments. If closing costs are financed, they are added to the base loan amount.

basement window inserts. N. The window frame and glass unit that is installed in the window buck, which is a square or rectangular box installed within a concrete foundation.

base rent. N. Monthly fixed rental payment not including utilities.

base shoe. N. Molding used next to the floor on an interior baseboard. Synonymous with *carpet strip*.

basis. N. The amount assigned to an asset from which gain or loss is determined for income tax purposes when the asset is sold. For assets acquired by purchase, basis is cost. Special rules govern the basis of property received by virtue of another's death or by gift, the basis of stock received on a transfer of property to a controlled corporation, the basis of the property transferred to the corporation, and the basis of property received upon the liquidation of a corporation.

basis point. N. Unit of measure for the change in interest rates. One basis point equates to one percent of one percent or 0.01.

bat. N. A half brick.

batt. N. A section of fiberglass or rock wool insulation.

batten. N. Narrow strips of wood used to cover joints or as decorative vertical members over plywood or wide boards.

bauhaus style. N. Houses in this German style of architecture had no decoration and were smooth white blocks constructed of steel and reinforced concrete with integral walls of glass.

bay window. N. A window that projects outward from the wall of a building.

beam. N. A long piece of wood or steel that supports a heavy building load. Synonymous with girder.

bearer. N. The legal owner of a piece of property.

bearing header. N. A beam placed perpendicular to joists and to which joists are nailed in framing for a chimney, stairway, or other opening.

bearing partition. See *bearing wall*.

bearing point. N. A point where a bearing or structural weight is concentrated and transferred to the building's foundation.

bearing wall. N. A wall that supports any vertical load in addition to its own weight. Synonymous with *bearing partition, load-bearing wall*.

bedrock. N. The layer of earth and stone suitable to support a structure.

before-and-after rule. N. In the case of eminent domain, a property to be taken by the government is appraised before and after the taking.

before-tax cash flow. N. Cash flow before income tax payments or adding income tax benefits.

before-tax income. N. Total income prior to deduction of taxes.

below grade. ADJ. Any portion of a project that is located below the ground level or below the surface grade of the surrounding land.

below the line. ADJ. Out-of-the-ordinary, non-recurring revenues or charges.

beneficial interest. N. Unit of ownership in a real estate investment trust.

beneficial use. N. Right of peaceful enjoyment of property by one party while another holds the legal title. This is applicable when rental property is involved.

beneficiary. N. One who inherits the value of insurance policies and investment funds after the owner dies.

bequeath. V. To pass property or personal effects to another through a will.

bequest. N. A gift by will of personal property. A bequest is not includable in the income of the recipient. Basis is usually set at the value of the property at the decedent's death. If a bequest of money is to be paid at intervals through property, it is taxable income to the recipient.

biannual. ADJ. Happens twice a year. Synonymous with *semiannual*.

bid. N. Stated price given to perform a job.

bidding requirements. N. The procedures and conditions for the submission of bids.

bidding war. N. Multiple offers to purchase a piece of property or competition between realtors for the listing of a piece of property.

bid out. N. A price obtained by contractors from subcontractors in order to estimate the building cost of a house or project prior to construction.

bid security. N. Funds or a bid bond submitted with a bid as a guarantee to the recipient of the bid that the contractor, if awarded the contract, will execute the contract in accordance with the bidding requirements of the contract documents.

bid shopping. N. When a contractor attempts to obtain prices from potential subcontractors and material suppliers that are lower than the contractor's original estimates on which their bids are based.

biennial. ADJ. Occurring every two years.

bifold door. N. Door composed of panels that are hinged vertically in the middle, which fold back upon themselves.

bilateral contract. N. A reciprocal contract in which the parties involved give mutual promises.

bi-level. N. A house that is built on two levels with the garage and storage or recreation room in the lower level and the balance of the house in the upper level. Homes of this style were built after 1950.

Bi-level

billing cycle. N. The period between billings for products and services, usually a month.

bill of assurance. N. A written guarantee of certain provisions, such as the size of individual homes, architectural standards, and quality of materials.

binder. N. (1) An agreement prior to contract between buyer and seller. (2) The report that is issued by a title insurance company detailing conditions of a home's title and giving guidelines for the title insurance policy.

binding arbitration. N. The judgment made by an independent third party to settle a dispute between two other parties; it may be either voluntary or compulsory.

biweekly payment mortgage. N. A mortgage that requires payments to reduce the debt every two weeks (instead of the standard monthly payment schedule). The twenty-six (or possibly twenty-seven) biweekly payments are each equal to one-half of the monthly payment that would be required if the loan were a standard thirty-year, fixed rate mortgage, and they are usually drafted from the borrower's bank account. The result for the borrower is a substantial savings in interest. Borrowers can qualify for a thirty-year monthly payment amount, but get a loan that pays off in approximately twenty-two years at current interest rates.

blanket insurance policy. N. Insurance policy covering multiple persons or pieces of property.

blanket mortgage. N. A mortgage that covers more than one parcel of real estate.

blankets. N. Fiberglass or rock wool insulation that comes in long rolls fifteen or twenty-three inches wide.

blended loan. N. Mortgage refinancing in which the new interest rate takes into account the interest rate on the prior loan and the prevailing current market rate.

blended rate. N. Interest rate of a blended loan, which exceeds the rate on the old loan but is less than the rate on new loans.

blind entry. A bookkeeping entry that shows debits and credits but neglects to record other essential information.

blind pool. N. A limited partnership in which the specific properties are chosen by the general partner after the funds become available.

blind pool syndicate. N. Money raised by a promoter and placed into a fund prior to the selection of investment properties.

blind trust. N. A financial situation in which a person appoints a fiduciary to manage his or her finances; used in cases where a conflict of interest is possible.

blockbusting. N. The illegal act of inducing homeowners to sell their homes or properties by representing that persons of another race are entering the neighborhood.

blown-in insulation. N. Insulation that is inserted into walls or other areas by being "blown in." Commonly used in attic floors or other areas that are inaccessible for normal installation.

blueprint. (1) N. A type of copying method often used for architectural drawings. (2) N. The drawing of a structure that is prepared by an architect or designer for the purpose of design and planning, estimating, securing permits, and actual construction.

Blue Sky laws. N. State statutes that protect the public against securities frauds of real estate companies. These include regulations over the licensing of brokers, registrations, new securities, and formal approvals by applicable government agencies.

blue stake. N. When a utility company comes and puts little flags in the ground to show where their service is located underground.

board foot. N. A unit of measure for lumber equal to 1 inch thick by 12 inches wide by 12 inches long. Examples: 1" x 12" x 16' = 16 board feet, 2" x 12" x 16' = 32 board feet

board of appeals. N. Governmental body that reviews property tax assessments procedures.

board of equalization. N. State board that ensures that local property taxes are assessed uniformly.

Board of Realtors®. N. Local group of real estate brokers who are members of the State and National Board of Realtors. They meet regularly to help determine licensing requirements as well as managing the multiple listing service of their area.

board of trustees. N. An appointed or elected body overseeing the management of an organization and rendering advice on issues, and is legally responsible for its decisions.

boilerplate. N. Form language used in legal papers, such as deeds and mortgages, before they are individualized with personal details.

bona fide. ADJ. (*Latin*) Persons or actions that are in good faith and honest.

bona fide purchaser. N. Buyer who is acting in good faith.

bond. N. (1) Agreement insuring one party against loss by actions or defaults of another. (2) An amount of money that legally should be on deposit with a government agency to secure a contractor's license.

bond for title. N. A property sales contract that is mutually binding on both parties, where the title remains with the seller until the buyer pays the purchase price; contract to convey title once certain contract terms are satisfied.

bonus room. N. A room with no designated function that can be used in many different ways, i.e., a kitchen or bedroom.

bookkeeping. N. The systematic recording of a company's financial transactions.

book profit. See *paper profit*.

books. N. A company's accounting records, such as ledgers and journals.

book value. N. The value of an asset as it appears on a balance sheet, equal to cost minus accumulated depreciation. Book value often differs substantially from market price, especially in knowledge industries such as the high-tech industry.

boom. N. A truck used to hoist heavy material up and into place at a construction site.

boot. N. Cash or property of a type not included in the definition of a nontaxable exchange. The receipt of boot will cause an otherwise tax-free transfer to become taxable to the extent of the lesser of the fair market value of the boot or the realized gain on the transfer. Examples of nontaxable exchanges that could be partially or completely taxable due to the receipt of boot include transfers to controlled corporations and like-kind exchanges.

borough. N. A section of a city having authority over local matters.

borrower. N. A person who receives money from a lender to buy property in exchange for a written promise to repay that money with interest.

borrower risk. N. Liabilities assumed by a borrower, such as loss of financial ability to repay a mortgage loan, rising interest rates on adjustable rate loans, and home devaluation.

bottom line. See *net profit*.

BPI. ABBRV. Buying Power Index.

breach. N. A violation of a law or obligation through commission or omission, where the responsibilities of an agreement or guarantee are not met. V. *breach*.

breach of contract. N. Failure to fulfill the terms of a contract, without legal, excusable reasons.

breach of covenant. N. Failure to fulfill a legal agreement.

breach of warranty. N. Inability, on the part of the seller, to pass along clear title to a buyer.

break-even point. N. Where total revenue equals total costs and there is no profit or loss, such as when an owner's rental income matches expenses and debt.

bridge loan. N. A short-term loan that allows a homeowner to purchase a home before selling the former residence. Synonymous with *gap loan, interim financing, standby loan, swing loan*.

broker. N. A person who acts as a conduit between two parties. A real estate broker is licensed to handle property transactions. A mortgage broker matches, for a specific fee, borrowers to lenders and loan programs.

brokerage. N. (1) The bringing together of two parties in exchange for a fee or commission. (2) A company or firm employing agents acting as brokers.

broker agreement. N. Contract to act on behalf of a principal in selling real estate, wherein the principal agrees to pay a commission to the broker when a buyer is produced who is ready, willing, and able to meet the terms of the sale.

brownstone. N. A nineteenth-century house, usually having four to five stories with a long staircase leading to the first floor. They share common walls with other brownstones.

Brownstone

budget. N. An itemized forecast of an individual's or company's income and expenses expected for some period in the future.

budgeting. N. Estimation of all income and expenses for an accounting period or financial forecasting, planning, and controlling.

buffer strip or zone. N. A piece of land separating two or more properties from each other.

buildability. N. Whether or not a home or other structure can be constructed on a piece of land.

builder breakdown method. N. Complete estimate of all costs in construction, including, but not limited to, land acquisition, material, labor, and preparation.

builder's risk insurance. N. Insurance coverage on a construction project during the construction phase, which may include additional coverage for the customer's sake.

builder upgrades. N. Better material or extras that are offered by the builder to the purchaser.

builder warranty. N. Limited-time warranty (one or two years) against defect, which is offered by builders to new homebuyers.

building capitalization rate. N. Rate of return of capital invested in building improvements.

building code. N. (1) Laws that control the construction or remodeling of homes or other structures. (2) Municipal or state ordinance that is enforceable under the police powers of the state and locality controlling alterations, construction methods and materials, size and setback requirements, use, and occupancy of all structures. Building codes have specific regulations covering all aspects of construction, and they are designed to maximize the health and welfare of the residents.

building density. N. Concentration of buildings in a given geographic area.

building inspector. N. An employee of a city or county who enforces building codes and ensures that construction is being performed correctly.

building life. N. Estimated useful life of a building.

building line. N. (1) Guidelines limiting the distance from the street or an adjacent property that a home or structure can be erected. Line established by a building or zoning code beyond which a building structure may not extend. (2) The outer edge of the rafter plate.

building loan agreement. See *construction loan*.

building moratorium. N. An enforced halt to construction, used to slow the rate of development in a town.

building permit. N. A permit, issued by local government, allowing construction or renovation of a house or other structure.

building residual technique. N. Appraisal method for determining a building's value on the basis of the residual building income after adjusting for land value.

building restrictions. N. Regulations limiting the type of use allowed on a piece of property.

building setback. N. Municipal ordinance stating the distance from a curb or property line where a building can be located.

building sewer. See *private sewer*.

build to suit. N. Where a landowner offers to build a specific structure on a site for a potential tenant and then lease the land and building to that tenant. This is done primarily for commercial property.

built-ins. N. Appliances or other items that are permanently attached to and framed into a home.

built-up roof. N. A roof composed of three to five layers of asphalt felt laminated with coal tar, pitch, or asphalt.

bullet loan. N. Intermediate debt (five to ten years) without periodic payments but with the entire amount (i.e., balloon payment) due at the maturity date.

bull nose. N. Rounded drywall corner.

bundle of rights. N. Various interests or rights an owner has in a property.

bungalow. N. A one-story house or cottage, which often has either an open or enclosed front porch.

Bungalow

Bureau of Land Management. N. The federal agency that oversees management of much of the government's undeveloped land.

burned-out tax shelter. N. An aging tax shelter where depreciation deductions have grown smaller over time.

burn rate. N. For a company with negative cash flow, the rate of that negative cash flow, usually per month.

business property energy tax credit. N. An energy tax credit allowed for the purchase of certain business-use property utilizing solar, geothermal, or biomass energy.

business-use property. N. Property used for the production of income. Examples include rental houses, machinery, factories, office buildings, and similar items.

buy-back. N. An agreement to sell real estate with a prearranged agreement to reverse the deal at an established price.

buydown. N. The payment of extra money on a loan, such as the payment of additional principal when making a monthly mortgage payment, to reduce the interest rate or life of the loan.

buydown mortgage. N. Home loan where the lender receives an up-front premium payment and, in return, reduces the interest rate during the early years of a mortgage.

buyer costs. N. There are other costs associated with the closing besides the down payment that are typically paid by the buyer.

They often include lender fees, advance payments or prepaid expenses, escrow accounts or reserves, title charges, recording and transfer fees, adjustments, and additional charges required by the lender.

buyer's broker. N. The real estate broker who represents only the buyer's interests in a transaction and whose commission is paid by either the buyer or through the seller or listing broker, at closing.

buyer's market. N. A real estate market in which the buyers have the advantage over the sellers due to slow sales.

buying power. N. Financial ability of a business or individual to afford a purchase, or the worth of the dollar in real terms considering inflation.

Buying Power Index. N. Average of income, retail revenue, and population of a locality as a percentage of the entire United States, which reflects the economic status of that region. ABBRV. *BPI.*

buy-out estimate. N. Estimated price at which one partner in a partnership can buy out another partner. Market comparisons, appraisals, or multiyear projections of market appreciation can be used to develop a buy-out amount.

buy-sell agreement. N. An agreement where the partners consent to purchase the interest of those leaving the partnership and those leaving consent to sell their interests to the other partners. Important in the event of death or disability of one or more partners, as well as other occurrences.

bylaws. N. Homeowners' association rules and regulations governing activities in certain types of home communities, condominiums, and townhouses.

bypass doors. N. Doors that slide by each other and are commonly used as closet doors.

cadastral map. N. Map within a jurisdiction that shows the boundary lines and ownership of all real estate in the area.

CADD. ABBRV. Computer-aided design and drafting.

CAE. ABBRV. Certified assessment evaluator.

CAGR. ABBRV. Compound annual growth rate.

caisson. N. A ten- to twelve-inch diameter hole drilled into the earth and embedded into bedrock at least three feet; the structural support for a type of foundation wall, porch, patio, monopost, or other structure. Two or more reinforcing bars (rebars) are inserted into and run the full length of the hole and concrete is poured into the caisson hole.

calendar year. N. A year that ends on December 31.

California bungalow. A small, one-story, early twentieth-century house.

California ranch. Long, one-story house that has a sloping roof with skylights and contemporary windows.

call option. N. Loan agreement clause allowing the lender to ask for the balance due at any time.

cancellation clause. N. The details under which each party may terminate an agreement.

cantilever. N. A projecting structure, such as a fireplace location or a bay window, where one floor extends beyond and over a foundation wall.

cap. N. A consumer safeguard that limits the amount the interest rate on an adjustable rate mortgage can change in an adjustment interval or over the life of the loan. For example, if a per-period cap is 1% and a borrower's current rate is 5%, his or her newly adjusted rate must fall between 4% and 6% no matter what the change is in the index the lender uses to set rates.

capacity. N. The borrower's ability to pay debt he or she incurs. Lenders base their evaluation on employment and wage information as well as credit history. Much of this latter data is available electronically through each borrower's credit report.

Cape Cod style. N. A style of wood-frame house with a steep roof and windows projecting outward from the second floor. Often, bedrooms are often on the first floor and there is a finished attic.

Cape Cod style

capital. N. (1) Money used to create income, either as an investment in a business or an income property. (2) The money or property comprising the wealth owned or used by a person or business enterprise. (3) The accumulated wealth of a person or business. (4) The net worth of a business, represented by the amount by which its assets exceed its liabilities. When lenders review the "capital" portion of a borrower's loan application, they review whether he or she has enough cash for a down payment and to pay closing costs. It is generally recommended that borrowers do not spend every last dime on the home purchase because there will be a significant need for cash after the borrower moves in.

capital appreciation. N. An increase in the market price of an asset. Synonymous with *capital growth*.

capital assets. N. Assets purchased for use over long periods of time, such as land and buildings, rather than for resale; these can be fixed assets consisting of tangible assets, such as plants and equipment, and intangible assets, such as patents.

capital expenditure. N. Money spent on improvements on a property, which becomes part of the cost of an existing fixed asset. The expense generally may be depreciated or amortized.

capital expenses. N. Amounts spent to buy or upgrade physical assets such as buildings and machinery. Synonymous with capital spending.

capital gain. N. Investment profit made from the sale of investments or real estate.

capital gain distributions. N. Amounts paid by mutual funds, regulated investment companies, and real estate investment trusts. These amounts represent the shareholder's portion of gain from the sale of capital assets owned by these investment companies. Capital gain distributions are taxed in the year constructively received and are always considered to be held long-term.

capital gain or loss holding period. N. The length of time the taxpayer owns a capital asset. Assets owned twelve months or less are held short-term; those owned more than twelve months are held long-term.

capital gains tax. N. Taxes placed on profits from the sale of investments or real estate.

capital growth. See *capital appreciation*.

capital improvement. N. An improvement made to boost the useful life of a property or add to its value. Major repairs such as the replacement of a roof are capital improvements. The costs of capital improvements to business property must be capitalized and may be depreciated.

capital investment. N. The money paid to purchase a capital asset or a fixed asset.

capitalization. N. Mathematical formula used by investors to compute the value of a property based on net income.

capitalization rate. N. The percentage rate of return estimated from the net income of a piece of property. Value = annual income ÷ capitalization rate. Synonymous with *income yield*.

capitalize. V. (1) To classify a cost as a long-term investment rather than charging it to current operations. (2) To record an expenditure having a benefit of more than one year to the cost of a property such as a new kitchen or a new roof. Upon sale of the property, the gain or loss, for tax purposes, is the difference between the selling price and the adjusted cost basis. If used for business, depreciation on the capital improvements may be deductible for tax purposes.

capitalized cost. N. Equipment valuation used in depreciation calculations.

capitalized interest. N. Interest that is not immediately expensed, but instead is treated as an asset and amortized over time in the income statement.

capital lease. N. Rental where the lessee obtains major property rights from the lessor, although not legally a purchase of the property.

capital liability. N. An obligation used to purchase fixed assets or to fund a specific project.

capital loss. N. The loss from the sale or exchange of a capital asset. Based on current law, up to $3,000 of net capital loss is deductible annually with the excess carried forward to future years. Losses on personal-use assets are not deductible.

capital net worth. See *net worth*.

capital recapture. N. Return of invested principal, excluding earned income or collection of a previously written off bad debt.

capital recovery. N. Amount of an investment, made in real estate, that is recovered.

capital resource. N. Any asset used in the production of products or services.

capital spending. See *capital expenses*.

capital turnover. N. Number of times a given amount of capital assets turn over to generate sales over a given period of time.

capped rate. N. Maximum interest rates a borrower might pay for an adjustable rate mortgage.

carpet strip. See *base shoe*.

carryback. N. Provisions in the tax code that allow certain losses or credits to be used in a prior year to the taxable year.

carryback financing. N. Financing of a property where the seller of the property holds a note for a set amount.

carrying charges. N. The cost of rental property to a landlord when that property is being rented or is vacant.

carryover. N. Provisions in the tax code that allow certain losses or credits to be used in a future tax year to the taxable year.

carryover basis. N. Basis for the valuation of property; for tax purposes, acquired from a decedent.

carve-outs. N. Items that a lender will require the borrower to personally guarantee for the life of the loan, which can include environmental problems, fraud, theft, and misappropriation of funds.

casement. N. Frames of wood or metal enclosing at least part of a window sash.

casement window. N. A window that is hinged on the side, which allows it to swing open outward. A quadrant gear forces a lever to open and close the window when the crankshaft, which is attached to a gear that turns the quadrant gear, is turned. Windows will be held in any position by the gearing, which can also be operated by remote control.

cash. N. Currency and coins on hand, bank balances, and negotiable money orders and checks.

cash accounting. N. The practice of keeping records of money received or expended.

cash asset ratio. See *cash ratio*.

cash basis. N. A cash-based transaction. Generally, cash basis bookkeeping is simpler than accrual basis bookkeeping, but makes securing financing more difficult.

cashbook. N. An accounting book that documents both cash receipts and disbursements.

cash budget. N. A forecast of estimated cash receipts and disbursements for a specified period of time.

cash control. N. The procedures used to verify the accuracy of cash receipts and disbursements.

cash conversion cycle. See *cash cycle*.

cash cycle. N. The length of time between the purchase of raw materials and the collection of accounts receivable generated in the sale of the final product. Synonymous with *cash conversion cycle*.

cash earnings. N. Cash revenues minus cash expenses. This differs from earnings in that it does not include noncash expenses such as depreciation.

cash equity. N. Amount, in cash, invested in property.

cash equivalency. N. Price for which real estate would be sold if all cash was involved.

cash equivalent doctrine. N. Generally, a cash-basis taxpayer does not report income until cash is constructively or actually received. Under the cash equivalent doctrine, cash-basis taxpayers are required to report income if the equivalent of cash (property, for example) is received in a taxable transaction.

cash flow. N. Cash remaining on a rental property when the operating expenses and loan payments are deducted from the gross rental.

cash flow statement. N. Financial records of receipts and expenditures during a specific period; a summary of a company's cash flow over a given period of time.

cashier's check. N. A check guaranteed by bank funds rather than a depositor's account.

cash journal. N. A journal where all transactions are initially recorded.

cash method of accounting. N. One of the two most common methods of accounting, the other being the accrual method. Under the cash method of accounting, income is reported in the tax year actually or constructively received and expenses are deducted in the tax year paid.

cash out. N. Any cash received at the time of a new loan beyond the balance of the mortgage; awarded based on the existing equity in the house. The cash out amount is calculated by subtracting the sum of the old loan and fees from the new mortgage loan and is available for all types of property.

cash-out refinance. N. A refinance transaction in which the amount of money received from the new loan exceeds the total amount of money needed to repay the existing first mortgage, closing costs, points, and to satisfy any outstanding subordinate mortgage liens. In other words, a refinance transaction in which the borrower receives additional cash that can be used for any purpose.

cash ratio. N. Total dollar value of cash and marketable securities divided by current liabilities. Synonymous with *liquidity ratio*, *cash asset ratio*.

casing. N. Wood trim molding installed around a door or window opening.

casualty. N. The complete or partial destruction of property resulting from an identifiable event of a sudden, unexpected, or unusual nature.

casualty loss. N. A tax deduction individuals may take if the loss is incurred in a trade or business, in a transaction entered into for profit, or if it is a personal loss arising from a disaster such as those that qualify casualties as such.

caveat emptor. N. (*Latin*) Let the buyer beware; the prospective buyer must examine the property thoroughly and buy at his or her own risk.

CD. ABBRV. Certificate of deposit.

CD-indexed ARMs. N. The Certificate of Deposit Index represents the weekly average of secondary market interest rates on six-month negotiable certificate of deposits. The initial interest rate and payments adjust every six months after an initial six-month period. Adjustable rate mortgages with this index typically come with a per-adjustment cap of 1% and a lifetime cap rate of 6%.

cease and desist order. N. Judicial order prohibiting a person or business from doing something, usually issued by the court when unlawful activity is occurring.

cease and desist petition. N. A notice filed by a homeowner, which notifies the secretary of state that a certain premises is not for sale and puts real estate brokers on notice that the premises has no implied invitation to be solicited.

ceiling. N. The maximum allowable interest rate of an adjustable rate mortgage.

ceiling joist. N. Framing used to support ceiling loads and supported in turn by larger beams, girders, or bearing walls. Synonymous with *roof joist*.

ceiling rate. N. A controlled or administered price, which is set for property by a federal or local agency, usually in extraordinary circumstances.

census tract. N. A geographic area mapped out by the U.S. government that includes demographic data that may be of interest to developers and other businesses.

CERCLA. ABBRV. Comprehensive Environmental Response, Compensation, and Liability Act.

certificate. N. Official, written documentation certifying that the fulfillment of certain requirements has occurred on a certain date.

certificate of beneficial interest. N. Document stating that someone has an ownership interest, but not direct control, in an asset, business, or estate.

certificate of deposit. N. Document representing that the bearer has a specified amount of money on deposit in a financial institution. ABBRV. *CD*.

Certificate of Deposit Index. N. An index based on the interest rates on six-month certificate of deposits; it is often used to determine the interest rate on some adjustable rate mortgages.

Certificate of Eligibility. N. Veterans Administration document verifying the eligibility of a veteran for a loan program.

certificate of occupancy. N. A document presented by a local government building department or other relevant agency that certifies a building or a specific part of a building has been inspected and approved for occupancy by tenants or owners. ABBRV. *CO*.

certificate of sale. N. Document issued by the court at a judicial sale, entitling the purchaser to receive a deed once the court approves the purchase.

certificate of title. N. Written opinion of the status of title to a property, given by an attorney or title company. This certificate does not offer the protection given by title insurance.

Certificate of Veteran Status. N. Federal Housing Administration form filled out by the Veterans Administration to establish a borrower's eligibility for an FHA veterans loan.

certification. N. Written statement of the correctness and reliability of something; written permission to do something.

certified assessment evaluator. N. Credential awarded by the International Association of Assessing Officers to appraisers of real estate working for a government body. ABBRV. *CAE*.

certified historic structure. N. A structure listed on the National Register of Historic Places or located in a designated historic area. The Internal Revenue Service Code provides tax incentives for the rehabilitation of such structures.

certified public accountant. N. An individual who has received state certification to practice accounting. ABBRV. *CPA*.

cession deed. N. Deed used to transfer property rights to a governmental authority.

CFA. N. Chartered financial analyst.

chain of title. N. The chronological order of conveyance of a property from the original owner to the present owner. Must be established before a lender will award a loan to a borrower.

change frequency. N. Adjustment schedule of an adjustable rate mortgage.

change in accounting method. N. A change from one method to another, which usually requires prior approval from the Internal Revenue Service. A change generally requires adjustments to avoid omissions or duplications.

change in accounting period. N. A change from one period to another. Income for the short period created by the change must be annualized to calculate the tax for that period.

change order. N. Any modification of a construction contract, which is signed by the owner, the architect, and the contractor, and would authorize a change in work, the amount of payment in the contract, or a change in the contract time.

Chapter 7. N. The most common form of bankruptcy for individuals. With this type of bankruptcy, most debts are wiped out. However, state law becomes very important on the issue of whether individuals can keep real estate. Florida, for instance, allows filers to keep their homes. Others only allow a small amount of home equity to be exempt. Once the filer has been legally cleared of debts, a Chapter 7 bankruptcy stays on a credit record for ten years. Eventually, filers may qualify for new consumer loans, but they will never qualify for the best rates.

Chapter 11. N. Type of bankruptcy filing allowing restructuring and reorganization of existing debts, which is used most often by businesses. Creditors must vote on a debt-paying plan and a judge must approve.

Chapter 13. N. Also known as reorganization, the Chapter 13 debtor agrees to a repayment plan over a number of years. The real estate advantage is that Chapter 13 filers are allowed to retain some secured debt, like a home or car. Also, a Chapter 13 bankruptcy will be listed on the debtor's credit report for seven years instead of ten.

charge off. See *bad debt*.

chartered financial analyst. N. An individual who has passed certain tests administered by the Institute of Chartered Financial Analysts of the Association for Investment Management and Research, including tests on economics, accounting, security analysis, and money management.

chart of accounts. N. A list of all account names and numbers used in a company's general ledger.

circuit breaker. N. A switch located on an electrical breaker panel or circuit breaker box.

city plan. N. Large-scale map of an urban area detailing land use; essential for projecting the growth and development of the urban area.

civil action. N. Legal proceeding instituted, by one party, to exercise a right in a disagreement between individuals or businesses.

civil court. N. State court where civil disagreements are decided by a judge or jury.

civil engineering. N. Specialization in the design of structures; buildings, bridges, etc.

civil law. N. Law involving noncriminal issues such as breach of contract, libel, accidents, etc.

class A, B, and C properties. N. Categories of various commercial properties based on their attractiveness—and pricing—in the marketplace. Class A buildings are well-designed and situated, and sought after by top tenants. Class A buildings are one step down from trophy buildings, considered the most desirable of all commercial properties. Class B buildings are clean spaces without many amenities. They lack architectural flair and are not in the most desirable locations. Class C buildings are usually unfinished spaces with only the basics in utilities and security, and are a draw for low-to-moderate income tenants who need affordable space and are willing to invest sweat equity in the project.

class action. N. Lawsuit brought by one or more persons of a large group for the benefit of all members of the group.

classified property tax. N. Tax rate that varies depending on the usage of the property in question.

cleaning deposit. N. Nonrefundable fee to pay for the painting and cleaning of an apartment or office after a tenant vacates the premises.

clear title. N. Title to property without liens, defects, or legal encumbrances of any kind. Synonymous with *good and marketable title*.

closed-end lease. N. Lease with monthly payments over a given period of time with no charge when the lease expires. At the expiration of the lease, the lessor sells the leased property for a gain or a loss.

closed-end mortgage. N. Mortgage in which the collateralized property cannot be used as security for another loan.

closing. See *settlement*.

closing agent. N. Someone who coordinates the closing meeting and keeps track of the documents while the borrower and his or her representative sign the papers.

closing costs. N. Expenses (over and above the price of the property) incurred by buyers and sellers in transferring ownership of a property. Closing costs normally include an origination fee, an attorney's fee, taxes, an amount placed in escrow, and charges for obtaining title insurance and a survey. Before a property can be transferred from one owner to another, closing costs must be paid. These costs may run anywhere from 3% to 6% of the total amount of the mortgage. Synonymous with *settlement fees*.

closing date. N. After a lender has approved a mortgage and the borrower accepts the commitment letter, the next step is to set a closing date. Many times, a real estate sales professional coordinates the setting of this date with the buyer, the closing agent, and the lender.

closing entry. N. The final bookkeeping entry made at the end of an accounting period to transfer income and expense items to the balance sheet accounts.

closing escrow. N. When all the conditions of the purchase and sale agreement have been fulfilled and the funds are paid out in accordance with the escrow agent's written summary of the funds received in escrow.

closing of title. N. Legal procedure in which property ownership is transferred.

closing statement. See *HUD-1 Settlement Statement*.

cloud on title. N. Any conditions revealed by a title search that adversely affect the title to real estate. Usually clouds on title cannot be removed except by a quitclaim deed, release, or court action.

CMA. ABBRV. Comparative market analysis.

CMO REIT. N. A real estate investment trust that invests in collateralized mortgage obligations.

CO. ABBRV. Certificate of Occupancy.

cobrokerage. N. Two or more authorized brokers who agree to cooperate together in representing a principal for the completion of a real estate sale.

code. N. An organized set of rules and regulations on a particular subject, which is often an accumulation of laws in a particular area of interest.

code of ethics. N. An organized group of ethical behavior guidelines that govern the day-to-day activities of a profession or organization.

codicil. N. An addition to a will; adding, subtracting, or clarifying provisions of the document.

COFI. ABBRV. Cost of Funds Index.

coinsurance. N. (1) Coverage involving the use of two or more insurers. (2) Arrangement where the insured and insurer share on a proportional basis the payment for a loss.

coinsurance clause. N. Provision in an insurance policy that caps the insurer's liability by stipulating that the owner of the property that has experienced damage must have another policy that covers usually 80% of the cash value of the property at the time of damage in order to collect the full amount insured.

collateral. N. Assets, such as a home, pledged as security for a debt.

collateral loan. N. Loan secured by the pledge of specific collateral, such as borrowing $20,000 against a savings account of $50,000.

collateral security. N. Additional security supplied by the borrower to obtain a loan.

collateralized mortgage obligations. N. A security backed by a pool of mortgage loans of various classes and maturities, and often packaged into real estate investment trusts. ABBRV. *CMO*.

collection. N. Series of steps by a lender taken to bring a delinquent mortgage current.

color of title. N. Unclear indications of ownership rights, which supplement a claim to title of property but do not actually establish it.

comaker. N. A person who signs a promissory note along with the borrower. A comaker's signature guarantees that the loan will be repaid, because the borrower and the comaker are equally responsible for the repayment. See also *endorser*.

combined financial statement. N. A financial statement covering multiple related or affiliated companies.

comfort letter. N. An accounting firm's statement, provided to a company preparing for a public offering, confirming that unaudited financial data in the prospectus follows generally accepted accounting principles and that no significant changes have occurred since the report was prepared.

commercial acre. N. An acre of property zoned for income-producing purposes.

commercial bank. N. A financial institution that provides business loans, credit cards, checking and saving accounts, etc. Commercial banks are the largest financial intermediaries directly involved in the financing of real estate.

commercial broker. N. Real estate broker who specializes in the listing and selling of commercial property such as businesses, industries, apartments, office buildings, etc.

commercial listing. N. List of business properties.

commercial property. Business property, such as an office building, medical center, hotel, store, etc., that is intended to operate with a profit.

commercial real estate. N. Real estate usable in a trade or business.

commingling. N. The mixing together of money, held in trust for one reason, with other money.

commission. N. Money paid to real estate agents on the sale of a home, usually a negotiable percentage of the sale price.

commission split. N. The method for dividing a commission between a registered real estate person and the sponsoring real estate broker and between the listing broker and the selling broker.

commitment. N. A formal offer by a lender stating the terms under which it agrees to lend money to a homebuyer. Also known as a "loan commitment." The commitment letter states the dollar amount of the loan being offered, the number of years a borrower has to repay the loan, the loan origination fee, the points, the annual percentage rate, and the monthly charges. The letter also states the time the borrower has to accept the loan offer and to close the loan. Synonymous with *loan commitment*.

commitment fee. N. The fee charged by the lender to guarantee that the commitment, with its terms intact, is available for a certain period of time.

committee deed. N. A deed in which two or more people in an indenture agreement have reciprocity and obligations toward each other.

common area. N. The area in a housing or condominium development that is owned by all residents.

common area assessments. N. Fees paid by a housing or condominium development, which are used to maintain, operate, or repair common areas.

common elements. N. The parts of a condominium that are owned by all of the unit owners.

common law. N. Laws based on custom, usage, and rulings of courts in various jurisdictions.

common-law state. N. A state in which the laws governing property rights are based on British common law. The property and income of each spouse belongs to him or her separately.

community association. N. A name given to any association of property owners sharing an interest in commonly owned property, which may be developed in condominiums, cooperatives, or housing subdivisions.

community income

community income. N. Income of a married couple, living in a community property state, which is considered to belong equally to each spouse, regardless of which spouse receives the income.

community property. N. A classification of property peculiar to certain states and referring to property accumulated through the efforts of both husband and wife.

community property laws. N. Statutes stipulating that the property accumulated during a marriage belongs equally to each spouse, regardless of how much each contributed.

Community Reinvestment Act. N. Law that encourages the loaning of money in neighborhoods where minority depositors live. ABBRV. *CRA*.

comortgager. N. Two or more parties sharing a joint financial obligation for a mortgage.

comparables. N. Recently sold properties that are used to determine the value of a similar property. See also *comparable sales*.

comparable sales. N. Method widely used by real estate brokers, where sales of similar properties in approximate neighborhoods are used to estimate property value.

comparative market analysis. N. The estimated value of property based on the comparison of similar properties. ABBRV. *CMA*.

comparative statements. N. Financial statements that follow a consistent format but cover different periods of time; useful for spotting trends.

competent parties. N. Persons considered legally capable of entering into a binding contract.

competitive bid. N. A bid that is made to ascertain the best bid for work to be done.

completion bond. N. Legal instrument used to guarantee the completion of a development according to specifications and is more encompassing than a performance bond, which assures that one party will perform under a contract on the condition that the other party performs. Assures production of the development without reference to any contract and without the requirement of payment to the developer.

compounding. N. The process of adding earned interest to the principal so that the interest is figured on a progressively larger amount; paying interest on interest.

compound interest. N. The interest paid on the principal amount plus interest that has accrued and been added to the total.

compound sum. N. Total amount, consisting of both principal and compound interest, due at maturity.

Comprehensive Environmental Response, Compensation, and Liability Act. N. See *Superfund*. ABBRV. *CERCLA*.

compressed buy-down. N. A buy-down mortgage where the level of rate reduction is changed every six months.

compressor. N. A mechanical device that pressurizes a gas in order to turn it into a liquid, thereby allowing heat to be removed or added.

comptroller. N. The chief accountant of a company. Synonymous with *controller*.

computer-aided design and drafting. N. A graphics platform used on computers so that drawing can be done in two or three dimensions; the three-dimensional designs are able to be animated to be viewed from different angles, including from the inside of the drawing. ABBRV. *CADD*.

concession. N. (1) Benefits granted by a seller/lessor to induce a sale/lease. (2) A right granted by a governmental body to use property for a particular type of business in a specific area.

condemnation. N. Process used by the government to take private property, without the consent of the owner, for the use of the public.

conditional commitment. N. A written promise by a lender to make a loan, pending certain conditions to be met by the borrower.

conditional conveyance. See *conditional sale*.

conditional offer. N. An offer to purchase a piece of real estate provided certain conditions are met.

conditional sale. N. A contract stating that the title will remain with the seller until certain conditions are fulfilled by the buyer.

conditional use permit. See *special use permit*.

condominium. N. A form of property, such as a building or complex, in which the homeowners hold title to individual dwelling units, and proportionate interest in common areas and facilities.

condominium conversion. N. Change in title from single ownership of an entire building to multiple owners of multiple units, i.e., a rental apartment house to individual condominium ownership. The tenant is usually given the right to purchase his or her unit at a favorable price, prior to open market sales.

Condominium Owners Association. N. An association that is comprised of the owners of condominium units and is concerned with the management of day-to-day matters in the complex.

conforming loan. N. A loan that meets the qualifications necessary to be purchased by Freddie Mac or Fannie Mae.

consideration. N. (1) An inducement, consisting of a thing that is legal and has value, for a person to enter into a contract. (2) The amount actually received from a sale after all expenses are deducted.

consolidated financial statement. N. A financial statement that covers a holding company and its subsidiaries.

consolidated metropolitan statistical area. N. An area with two or more primary metropolitan statistical areas. A consolidated metropolitan statistical area must also include at least one million people ABBRV. *CMSA*.

consolidation loan. N. Loan that combines smaller loans into one larger loan; typical of a refinance of debt.

construction contract. N. A legal document that outlines all the specifics and dollar amounts of a construction project. It should contain the contractor's registration number, a statement of work quality, a set of blueprints or plans, a construction timetable, a set of specifications, a fixed price for the work, a payment schedule, any allowances, a clause outlining how any disputes will be resolved, and a written warranty.

construction documents. N. Drawings and specifications from an architect that provide the detailed requirements of a construction project.

construction drawings. N. Includes architectural plans, building plans, working drawings, blueprints, etc. Construction drawings provide the information, drawings, and instruction needed to construct a building.

construction loan. N. Short-term loan used during the construction of a building or home. Funds are disbursed in stages, according to amount of work completed. Synonymous with *building loan agreement, development loan.*

construction management contract. N. Contract between an owner and the general contractor who is responsible for hiring and supervising the trades and tasks necessary to build a structure.

construction specifications. N. Commonly referred to as "specs" these detailed descriptions consist of all items necessary to complete the construction of a building, including drawings and blueprints that outline the technical standards.

construction-to-permanent loan. N. A construction loan that can be converted to a long-term traditional mortgage upon completion of the construction.

constructive eviction. N. The altering of rented or leased premises by a landlord, rendering it unsuitable for habitation in order to effectuate the tenant's vacating.

Consumer Price Index. N. The most widely known of many such measures of price levels and inflation that are reported to the U.S. government. It measures and compares, from month to month, the total cost of a statistically determined "typical market basket" of goods and services consumed by U.S. households. It is one of many economic statistics the Federal Reserve uses to set interest rates. ABBRV. *CPI.*

contingency. N. A condition that must be satisfied before a contract is legally binding. The most common contingencies include professional home inspection; inspection for termites, asbestos, formaldehyde, and lead-based paint; and, testing for radon and hazardous waste sites.

contingency clause. N. A condition that must be fulfilled in a purchase contract.

contingency fund. N. Money set aside for a possible loss.

contingency listing. N. A property listing with a special condition that must be met.

contingency reserve. N. Most mortgages for purchase-renovation require an additional 10% of the total cost of the project to be put aside into a reserve account. This contingency reserve is only used when unforeseen repairs or deficiencies are found during renovation.

contingent fee. N. Fee that must be paid upon the occurrence of certain events.

contingent sale. N. Sale that is finalized only in the case of a particular occurrence.

continuation statement. N. Document submitted to a government agency to extend the time period for a previously approved document.

continuity tester. N. A device that tells whether a circuit can carry electricity.

contract. N. An agreement between two or more parties to do something specific.

contract for deed. N. Contract where the seller agrees to defer all or part of the purchase price for a specified period of time.

contract for novation. N. In law, substituting a suitable person or entity for an original party of a contract, which terminates the old contract and begins a new one.

contract of sale. N. The written agreement between the buyer and seller on the purchase price, terms, and conditions of a sale.

contractor. N. A person or firm supplying materials or work, for a stipulated sum, in the building trade. Prime contractors are

contract price

responsible for the entire job as a whole, while subcontractors are responsible for a certain trade and contract with the prime contractor.

contract price. N. An amount payable to the seller and equal to the gross selling price when no mortgages are involved. If a mortgage is assumed, the contract price is the gross selling price minus the amount of the mortgage plus the excess (if any) of the mortgage over the seller's basis and expenses of sale.

contract specifications. N. Details of a contract of sale including a legal description, type of deed, closing information, etc.

contract to purchase. See *agreement of sale*.

contractual lien. N. Voluntary obligation or encumbrance, such as a mortgage.

controller. See *comptroller*.

conventional loan. N. Long-term loan made for the purchase of a home, which is not insured or guaranteed by a governmental agency and which generally conforms to the standards required for sale of the loan into the secondary mortgage market. Typically requires a substantial down payment, and is usually only available to those having good credit. It has fixed monthly payments for the life of the loan and usually has a fifteen-, twenty-, or thirty-year period of fixed interest rates.

conversion clause. N. A provision in some adjustable rate mortgages that allows a borrower to change an ARM to a fixed-rate loan, usually after the first adjustment period. The new fixed rate will be set at current rates, and there may be a charge for the conversion feature.

convertible adjustable rate mortgages. N. A mortgage that begins as an adjustable rate loan but can be converted to a fixed-rate mortgage during a specified period of time.

conveyance. N. The transfer of title of property from one person or entity to another.

conveyance tax. N. Tax imposed on the transfer of real estate.

co-op. See *cooperative*.

cooperative. N. Business owned and managed by its members, in which profits and costs are shared. Synonymous with *co-op*.

cooperative apartment. N. Apartment building in which each resident owns a percentage share of the corporation that owns the building.

cornice. See *overhang*.

corporate relocation. N. An arrangement in which employers pay for the transfer and move of employees.

corporation net worth. N. Total assets less total liabilities.

correction deed. N. Deed issued to correct errors made in another deed.

cosign. V. To sign a note for the benefit of another, therefore assuming liability for the debt.

cost accounting. N. The process of identifying and evaluating production costs.

cost depletion. N. A method for recovering the taxpayer's investment in natural resources or timber.

cost of capital. N. The rate of return that is necessary to maintain market value of a real estate project and is also used for project evaluation purposes.

cost of development. N. Expenditures incurred to develop real estate.

Cost of Funds Index. N. An index of the weighted average interest rate paid by savings institutions. ABBRV. *COFI*.

cost of maintaining a home. N. Expenses necessary to maintain a taxpayer's residence. These costs include rent or mortgage interest and real estate taxes, fire and casualty insurance on the dwelling, upkeep and repairs, utilities, paid domestic help, and food consumed in the home.

cost-plus percentage contract. N. A contract that determines the builder's profit based on a percentage of the labor and materials used in the construction of the building.

counterclaim. N. Counteraction by a defendant against a plaintiff in a legal action.

cove molding. N. A molding with a concave face used as trim or to finish interior corners.

covenant. N. A binding agreement made by two or more parties to either do or keep from doing a specified thing.

covenant running with the land. N. Written agreement or guarantee annexed to the land between two or more parties to do or not do something; this is transferred to successive titleholders.

CPI. ABBRV. Consumer Price Index.

craftsman style. N. Architectural style that evolved near the turn of the century as part of the Arts and Crafts Movement.

Craftsman style

crawl space. N. A shallow space below the living quarters of a house, which is normally enclosed by the foundation wall and has a dirt floor.

creative financing. N. Innovative financing arrangements to help sell a property.

credit. N. Money a lender extends to a borrower who gives a commitment to repay the loan within a certain amount of time.

credit application. N. Form used to record information about an applicant seeking a loan.

credit bureau. N. An agency that collects and sells individuals' credit information. Consumers can order a free copy of their credit report from one of the three main credit bureaus, i.e., Equifax, Experian, and TransUnion.

credit history. N. A record of an individual's open and repaid debts. A credit history helps a lender to determine whether a potential borrower has a history of repaying debts in a timely manner. When a lender reviews an applicant's credit history, the lender examines all the information in a credit report. This includes credit cards, student loans, automobile loans, and other loans. The lender reviews whether payments have been made on time or late. A credit report also indicates if creditors discharged a debt because they believed it would never be repaid, if the borrower declared bankruptcy, or if the borrower went through foreclosure.

credit inquiries. N. Every time a lender reviews a borrower's credit history, an inquiry is recorded in their credit report. Having many recent inquiries may suggest the use of credit is increasing, which is often viewed as risky for the lender. However, auto and mortgage loan inquiries in the thirty days prior to the score being calculated are not used.

credit life insurance. N. Insurance that pays off a mortgage in the event of the borrower's death.

credit limit. N. Maximum amount of money that can be loaned to a prospective borrower.

creditor. N. A person or institution to whom a debt is owed.

credit rating. N. Degree of creditworthiness assigned to a person based on credit history and financial status.

credit report. N. A report detailing the credit history of a prospective borrower that is used to help determine borrower creditworthiness. See also *credit scoring*.

credit report errors. N. Inaccuracies on a credit report. The best way to prevent or remove errors is to request copies of your credit report every year or before a major purchase and review the information. Since each of the three main credit bureaus (Equifax, Experian, and TransUnion) keeps their own records, it may be wise to request copies from all of them. Expect to pay a fee.

Correcting Your Credit Report

Under the Fair Credit Reporting Act, a borrower has the right to correct any errors found on a credit report. Here are steps to follow.

1. Tell the credit agency in writing that the information it has is inaccurate. The letter should clearly identify each disputed item in the report, state the facts, and enclose copies of any evidence supporting the claim. The borrower should explain why he or she disputes the information, and request a deletion or correction. Circle the relevant information on the credit report and attach.

2. Mail this information by certified mail to ensure the agency receives it.

At the same time, a borrower should write to the specific creditor he or she believes acted in error and include the same evidence sent to the agency. This could speed up action on the matter. Many creditors have specific addresses and phone numbers for disputes, so check that information so the case ends up in the right hands.

3. The credit agency must reinvestigate the items in question—usually within thirty days—unless it considers the dispute frivolous. It also must forward all relevant data you provide about the dispute to the information provider. After the information provider—the credit card company, utility, bank, etc.—receives notice of a dispute from the agency, it must investigate, review all relevant information provided by the agency, and report the results back to the agency.

4. If the information provider finds it did make an error, it must notify all credit agencies so they can correct the information. Disputed information that cannot be verified must be deleted from the borrower's file.

    Note: If a borrower was late with a payment but is now current with a particular account, the credit report must reflect it.

5. When the agency's reinvestigation is complete, it must provide the borrower with written results and a free copy of the revised credit report reflecting the change. If an item is changed or removed, the agency cannot put the disputed information back in a borrower's file unless the information provider verifies its accuracy and completeness and supplies contact numbers.

6. If a borrower requests, the agency is also required to send correction notices to anyone who received the borrower's report in the past six months. Job applicants can have a corrected copy of their report sent to anyone who received a copy during the past two years for employment purposes.

7. Lastly, most accurate negative elements of a credit report must be removed after seven years. The exceptions are:

- information about criminal convictions—no time limit;

- bankruptcy—depending on the type filed, bankruptcy can remain on a credit report for seven to ten years;

- credit information reported in response to an application for a job with an annual salary of more than $75,000—no time limit;

- credit information reported because of an application for more than $150,000 worth of credit or life insurance—no time limit; and,

- information about a lawsuit or an unpaid judgment—can be reported for seven years or until the state's statute of limitations runs out, whichever is longer.

credit report fee. N. Fee that covers the cost of the credit report, which the lender uses to determine the borrower's creditworthiness.

credit repository. N. A large company that gathers financial and credit information from various sources about individuals who have applied for credit.

credit score. N. A single number calculated from various credit report data that serves as a guide to lenders and other subscribers—including employers—on how creditworthy a person is. The scores used in mortgage lending are typically in the three hundred to nine hundred range; the higher the score, the better. A credit score is one of several measurements a lender will use to evaluate a loan application; others include credit history, credit inquiries, outstanding debt, payment history, and the types of credit a borrower uses.

credit union. N. Nonprofit cooperative organization providing banking and financial services such as home improvement loans, home equity loans, and mortgages to its members.

creditworthiness. N. When lenders judge borrowers to be worthy of future credit.

crown molding. N. A molding used on cornice or wherever an interior angle is to be covered, especially at the roof and wall corner.

cumulative. N. An arrangement in which a payment not made when due is carried over to the following period.

current assets. N. Assets that can be converted to cash in less than one year.

current capital. N. Current assets minus current liabilities; the part of a company's capital that is used in everyday operations.

current debt. N. See *current liabilities*.

current liabilities. N. Expenses that are due to be paid. Synonymous with *current liabilities*.

current ratio. N. Current assets divided by current liabilities; an indication of a company's ability to meet short-term debt obligations. The higher the ratio, the more liquid the company.

current value. N. Value of a home at the time of appraisal.

custom builder. N. A builder who constructs a home or building with plans selected by the owner.

custom-built. ADJ. Made according to an owner's specifications.

custom home. N. A structure designed by an architect selected by the owner.

dado. N. A groove cut into a board to receive a connecting board.

damage deposit. N. Prepayment required to cover damage by a tenant, other than normal wear and tear.

damper. N. A metal door placed within a fireplace chimney that can be closed when the fireplace is not in use.

dampproofing. The use of tar or another waterproof material on the exterior of a foundation wall.

days on the market. N. Period of time a property is listed for sale prior to being sold or removed.

days payable. N. A measure of the average time a company takes to pay vendors, equal to accounts payable divided by annual credit purchases times 365.

days receivable. N. A measure of the average time a company's customers take to pay for purchases, equal to accounts receivable divided by annual sales on credit times 365.

DBA. ABBRV. Doing business as.

dealer. N. A person or firm that regularly buys and sells property. A person is classified as a dealer if at the time of the sale that person held the property primarily for sale to customers in the ordinary course of business. Gains from the sale of such property are ordinary gains, not capital gains.

debenture. N. A long-term bond or note issued by governments or corporations and not secured by a mortgage or property.

debit. N. An accounting entry, which results in either an increase in assets or a decrease in liabilities or net worth.

debit note. N. A note indicating an amount owed by a person or company. Serves the same function as an invoice.

decimal feet

debt. N. Money or another item of value owed by an individual or an organization for a sum it has borrowed. Debt can be secured by a note, bond, or mortgage that states repayment terms and, if necessary, interest requirements.

debt financing. N. The raising of money by loans and borrowing directly from financial institutions, providing increased financial leverage. Interest may be tax deductible.

debt limit. N. Maximum amount of debt an individual or business can borrow.

debtor. N. Individual or entity owing money.

debt ratio. N. Debt capital divided by total capital.

debt service. N. The interest and principal paid on a loan.

debt service coverage ratio. N. In mathematical terms, the annual net operating income of a property divided by the annual debt service of the mortgage loan on the property. This ratio is used by lenders and investors to help them figure out whether the property in question will generate enough income to pay the mortgage. The higher the ratio, the better. Also known as the debt service ratio.

debt-to-income ratio. N. Ratio of monthly debt payments to monthly gross income. Lenders use housing debt-to-income (DTI) ratio (housing payment divided by monthly income) and a total DTI ratio (total debt payment—including the house payment—divided by monthly income) to determine whether a borrower's income qualifies a buyer for a mortgage.

decedent. N. One who has died with a valid will in effect.

decimal feet. N. Measurement of length in feet and decimal portions. For example, 5.5 in decimal feet is also 5 feet 6 inches or 5½ feet.

declaration of homestead. N. Statement filed with a government authority declaring the property a homestead for purpose of securing a homestead exemption; the declaration has no effect on the title and is not a conveyance.

declaratory judgment. N. A binding determination by the court as to whether there is an allowable action between the litigants. A subsequent trial determines relief.

declining balance depreciation. N. An accelerated method of depreciation. The type of property determines the percentage the depreciable basis for the next year is reduced by the depreciation deduction taken in the current year.

declining market. N. Market condition in which there are more sellers than buyers, causing prices to fall.

decree. See *judgment*.

dedication. N. Property given and accepted as a grant to the public.

deduction. N. An amount that may be subtracted from income that is otherwise taxable.

deed. N. Legal document transferring ownership of a piece of property from one owner to another. It contains a description of the property, and is signed, witnessed, and delivered to the buyer at closing.

deed covenant. N. Any of a number of types of covenants agreeing to do or not to do something that is attached to the title, such as the use of an architectural style or a type of material, and passes from one owner to the next.

deed description. N. Property description contained in a deed.

deed-in-lieu. N. A deed given by a mortgagor to the mortgagee to satisfy a debt and avoid foreclosure. Synonymous with *voluntary conveyance.*

deed in lieu of foreclosure. N. Legal document conveying property to the lender after the borrower defaults on his or her mortgage payment.

deed of confirmation. N. Often referred to as a correction deed; used to rectify errors made in a previous deed.

deed of release. N. The deed that releases property or a portion of it upon satisfaction of a mortgage or other debt.

deed of trust. N. A document that gives the lender the right to foreclose on a piece of property if the borrower defaults on the loan. In some states, a deed of trust is used instead of a mortgage. When homeowners sign a deed of trust, they receive title to the property but convey title to a neutral third party—called a trustee—until the loan balance is paid in full.

deed recording fee. N. Fee charged by the government to enter into the public record the deed and documents relative to the transfer of title to a piece of property.

deed restrictions. N. Written statements in a deed that outline the limits of use of a property. Restrictions that are imposed against the race, sex, nationality, color, or creed of a person are illegal.

de facto contract. N. Contract intended to convey property from one individual to another but that is defective in one respect, such as providing no consideration for the exchange.

default. N. Failure of a debtor to pay on a due date, or failure to fulfill a duty or discharge an obligation, such as a mortgage payment.

default charge. N. Penalty charged if the amount owed on a purchase of real estate is not paid on time.

default judgment. N. Judgment issued by the court against a defendant who does not respond to the plaintiff's lawsuit and does not respond in his or her own defense.

defeasance. N. When the lender replaces the cash flows of the original loan with securities, usually Treasury bonds. The borrower pays the lender enough money to buy these securities and the lender goes out in the bond market and buys the right combination of bonds. This allows the property to be released as collateral for the loan and the treasuries become the new loan collateral.

defeasance clause. N. Provision guaranteeing the return of title to a mortgagor upon satisfaction of a mortgage's condition and terms; causes the discharge of a mortgagee's estate interest in a property.

defeasible clause. N. Provision in a contract, title, or mortgage that is subject to be repealed or revoked upon the satisfaction of a claim or completion of a future event.

defeasible title. N. Title that can be made null and void or defeated upon the satisfaction of a claim or the completion of some future contingency.

defective title. N. A title obtained through error or fraud without proper signature or consideration, or because of another improper action. A defective title is null and void, having no effect on the original title.

defendant. N. In civil court, the individual against whom a court action is brought by a plaintiff for restitution of property or satisfaction of a complaint. In a criminal court, an individual accused of a crime.

deferred charge. N. An expenditure that is considered an asset until it becomes relevant to the business at hand, such as prepaid rent that is considered an asset until the rent is officially due.

deferred credit. N. Revenue received by a firm but not yet reported as income.

deferred interest mortgage. N. This mortgage has a lower interest rate and, thus, a lower monthly mortgage charge. When the house is sold, the lender receives the deferred interest plus a fee for postponing the interest that would have been paid monthly.

deferred maintenance. N. Postponed repairs or maintenance on a piece of property, which result in a decline of property value.

deferred payment. N. Money payment that is to be delayed until a future date or for an extended period of time.

deferred revenue. N. Revenue that is considered a liability until it becomes relevant to the business at hand, such as a payment received for work that has not yet been performed.

deferred tax. N. A liability that results from income that has already been earned for accounting purposes but not for tax purposes.

deficiency. N. Additional tax liability that the Internal Revenue Service deems to be owed by a taxpayer.

deficiency judgment. N. Court finding that the debtor owes an amount exceeding the value of the collateral put up for the defaulted loan.

deficit. N. The amount by which a company exceeds its budget. Synonymous with *deficit spending*.

deficit net worth. N. On a balance sheet, the excess of liabilities over assets and capital stock, usually resulting from operating losses. Synonymous with *negative net worth*.

deficit spending. See *deficit*.

delamination. N. Failure of a construction adhesive.

delinquency. N. A debt for which a payment is overdue.

delinquent mortgage. N. A mortgage involving a borrower who is behind on payments. If the borrower does not bring the mortgage to date within a specified amount of time, the lender may begin foreclosure proceedings.

delivery. N. Transfer of property from one entity to another.

delivery basis. N. Method of revenue recognition based on delivery instead of sale.

demand loan. N. Loan with no established maturity period, which is callable on the demand of the lender for repayment. The interest is calculated on a daily basis and paid periodically.

demise. V. (1) Transfer of an estate by bequest or contract for a stated time period or life. (2) The making of a charter or lease for a specified time period.

demised premises. N. Leased or rented property.

demising clause. N. Provision in a lease whereby the landlord leases and the tenant takes the property. See also *lessor, lessee*.

demising wall. N. A partition wall separating one tenant's space from another's, or from a common area.

demographics. N. Statistics based on family size, age, occupation, marital status, and other population characteristics. This data is used by developers, lenders, and other businesses to develop their products and services.

demolition insurance. N. Insurance policy that indemnifies the property owner up to the limits of the policy against fire or another hazard requiring the total destruction and removal of the structure.

demurrer. N. A legal motion by a defendant that says there are not enough facts to prosecute him or her or that the court has no jurisdiction to do so.

Department of Veterans Affairs. N. An agency of the federal government that guarantees residential mortgages made to eligible veterans of the military services. The guarantee protects the lender against loss and thus encourages lenders to make mortgages to veterans. ABBRV. *VA*.

deponent. N. One who acts as a witness and gives written testimony under oath.

deposit. N. Money given, along with an offer to purchase property or as security for the performance of some contract; intended to show willingness to follow through with the purchase agreement. Synonymous with *earnest money*.

deposition. N. Discovery of information before trial, where a stenographer records the statements made by a witness under oath. These statements are made to answer questions posed by the attorneys to both parties.

Depository Institutions Deregulation and Monetary Control Act. N. Federal law that represented significant decontrol of federally regulated banks and saving and loan associations. It removed interest rate limitations, authorized interest-bearing checking accounts, reduced the applicability of states' usury laws, and widely expanded the services of S&Ls.

depreciable life. N. Economic or physical life of a fixed asset.

depreciable real estate. N. Under current tax law, real estate is depreciated under either the straight-line method or the modified accelerated cost recovery system method.

depreciated cost. N. The original cost of an asset minus its total depreciation thus far. Synonymous with *net book value, written-down value*.

depreciation. N. A For tax purposes, the deduction of a reasonable allowance for the wear and tear of assets—including real estate—used in a trade or business or held for the production of income.

depreciation basis. N. Amount subject to depreciation, which equals the initial cost less the estimated salvage repair.

depreciation recapture. N. Part of a capital gain constituting tax benefits previously taken and taxed as ordinary income.

depressed market. N. Market condition in which the prices of real estate are declining due to lack of demand. Synonymous with *weak market*.

derivative title. N. Transfer of title based on a preceding title transfer. A derivative conveyance increases, moderates, renews, or transfers the stake created by the original conveyance.

derived demand. N. Secondary demand that is created because of a primary agent or facility, such as an office building creating a need for a coffee shop.

descriptive memorandum. N. A type of description of a real estate property offering by a developer rather than a prospectus.

design/build. N. Project where the owner contracts with a company to perform both design and construction services.

design drawings. N. Plans that are used in the building business that give, in detail, all needed information for the construction or fabrication of a building or structure.

designer. N. Home design professional. Designers are limited to drawing blueprints, unlike architects who are able to certify plans.

design load. N. Maximum amount of weight that can be supported by a structure.

desist and refrain order. N. Court order to stop a specific activity.

detached housing. N. Freestanding residential housing constructed on its own building lot.

developer. N. The person or company that builds new homes, shopping centers, or commercial buildings for a profit on a specific area of land. A developer will organize and plan the development, supervise its construction, and manage all the business elements of the project.

developer's equity. N. Financial interest a developer has in a project.

developer's profit. N. The sum of money a developer earns, after all costs are deducted, in a development project.

development. N. The planning and building of homes, shopping centers, business facilities, schools, churches, etc. The process includes the construction of streets, sewers, utilities, parks, etc. In some cases, it may simply describe the process of obtaining the required governmental approvals for construction to proceed. The phrase "developed land" usually describes property for which

development loan. N. See *construction loan*.

devise. N. Gift of real estate as stipulated in a will.

devisee. N. One who receives real estate under a will.

devisor. N. Someone who donates real estate.

dictum. N. Judge's remark that illustrates or amplifies the ruling. Also describes a ruling made by an arbitrator.

dimension plans. N. Initial plans that show the layout of a house but that are less detailed than full blueprints.

direct capitalization. N. Divides a property's first-year net operating income by an estimated general capitalization rate to develop a total property estimate. For example, if an income property produces a first-year net operating income of $30,000 and the market indicates a general capitalization rate of 10% for comparable properties, the direct capitalization estimate of the value of the total property would be $300,000 ($30,000 x 10).

direct costs. N. Site preparation or building construction costs, including fixtures. This does not include such costs as building permits and land surveys, or overhead costs such as insurance and payroll.

direct overhead. N. Cost of doing business on one specific job.

direct reduction mortgage. N. A fully amortized mortgage necessitating periodic payments of both interest and principal. In the early years of the loan, the share of principal is smaller and the interest larger. This gradually reverses toward the end of the period. ABBRV. *DRM*.

disclosure statement

disability insurance. N. Insurance policy that covers an individual's ability to produce income.

disaster loss. N. If a casualty is sustained in an area designated as a disaster area by the President of the United States, the casualty is designated a disaster loss. For tax purposes, a disaster loss may be applied to the previous year before the disaster occurred so the victims can get immediate benefits.

disbursement. N. Money paid out in the discharge of a debt or expense.

discharge. V. Removing a debt by making payment.

discharge of bankruptcy. N. Court order whereby the bankrupt debtor is forgiven of his or her debts. Depending on the form of bankruptcy, a bankruptcy remains on a credit report for seven to ten years.

discharge of lien. N. Order to withdraw a property lien after a claim is paid by other means.

disclaimer. N. Renunciation of a claim to real estate ownership.

disclosed principal. N. A principal-agent transaction or contract where a third party knows the name of the principal the agent represents. In this arrangement, the agent is not legally bound under the written or oral agreement.

disclosure statement. N. (1) A written statement of a borrower's rights under the Truth in Lending Act or a statement of all financing charges, which must be disclosed by a lender. (2) A statement that lists information relevant to a piece of property, such as the presence of radon or lead paint.

discount. N. The deduction of a certain amount from a payment based on negotiation or the introduction of new information that could lower the cost of the item.

discounted cash flow. N. A method to estimate the value of a real estate investment, which emphasizes after-tax cash flows and the return on the invested dollars discounted over time to reflect a discounted yield. The value of the real estate investment is the present worth of the future after-tax cash flows from the investment, discounted at the investor's desired rate of return.

discount loan. N. Loan in which the entire financing charge is subtracted from the initial loan proceeds. The total amount of funds received is the face value of the loan less this deduction. For example, with a $50,000 one-year loan borrowed at a discount rate of 12%, the amount disbursed at the loan closing would be $44,000. The effective interest rate would be 13.6%, not the 12% discount rate, since only $44,000 is received.

discount point. N. An additional fee on home mortgages, payable in cash at the time of closing. One point equals 1% of the loan amount; for a mortgage of $100,000, one point would equal $1,000. Typically, each point paid for a thirty-year loan lowers the interest rate by .125%. If the current interest rate on a thirty-year mortgage were 7.75%, paying one point would lower the interest rate to 7.625%. It often makes more sense to pay discount points up front if the borrower plans to stay in the home for a long time.

discount rate. N. The interest rate charged by the Federal Reserve Bank to its member banks for loans. Rate changes will have a significant impact on the real estate market.

discrimination. N. Unequal treatment and denial of opportunity to individuals based on race, color, creed, nationality, age, or sex.

disinflation. N. A lessening in the rate of inflation that may occur during a recession.

disposable income. N. Personal income minus personal income tax payments and other governmental deductions; the money available for people to spend or save.

dispossess proceedings. N. Legal action by the owner of property to oust or exclude an individual or business from using the property.

distraint. N. When tenant property is seized legally or illegally by a landlord.

distressed property. N. Property in poor financial or physical condition; i.e., foreclosed real estate, property in a bankruptcy, or income property that is making an inadequate return.

distribution approach. N. The apportioning, disbursing, dividing, or parceling out of property among individuals, e.g., in probate, the approval of the court to divide and distribute the contents of an estate after all claims against it are satisfied; the estate is then divided between all distributees.

divestiture. N. (1) The voluntarily or involuntarily surrender of ownership of property or an interest therein. (2) A court order to give up possession or the right to property such as in the case of an antitrust action.

doctrine. N. Legal rule, principle, or tenet.

document. N. Recorded materials including letters, photos, reproducible computer files, legal forms, etc. A document is any tangible information including letters, contracts, electronic or paper files, X-rays, receipts, or other material evidence.

documentary evidence

documentary evidence. N. Any written evidence or tangible material that is coherent and related to the subject at hand. This includes documents, contracts, electronic and paper files, photographs, and other nonoral evidence.

document needs list. N. List of documents that a lender requires from a potential borrower, such as paycheck stubs and credit card statements.

document stamp. N. Tax imposed by some state and local governments to record property deeds and mortgages in the public records.

doing business as. N. Certification by a state that a principal is doing business under an assumed name. The certification also contains the address where the business is being conducted. ABBRV. *DBA*.

dollar amount paid. N. Cash plus the principal amount of a loan on the property that the taxpayer is legally obligated to pay.

dollar and percentage adjustment. N. Modification in the amount of money involved for some justifiable reason.

dollar stop. N. An agreed amount of taxes and operating expenses for which the tenant will pay a prorated share of increases. It may apply to property taxes, insurance, or other expenses.

dominant tenement. N. Property that has an easement right through another adjoining property. The property through which the easement passes is considered to have the subservient tenement.

donee. N. One to whom a gift or bequest is made.

donor. N. One who gives a gift or bequest.

door. N. A swinging panel that closes off an entrance.

dormer. N. A window that projects from a sloped roof to expand livable space in an attic.

double budget. N. An accounting system that keeps capital expenses and operating expenses separate.

double-declining balance depreciation. N. An accelerated depreciation method in which a fixed percentage factor of two times the straight-line rate is multiplied each year by the declining balance of the fixed asset's book value. To compute the annual depreciation expense, the asset's book value at the beginning of the period is multiplied by the double declining rate. Although salvage value is not included in the initial calculation for depreciation, a fixed asset cannot be depreciated in the last year below its salvage value.

double-digit inflation. N. Annual rate of inflation of 10% or higher.

double-entry bookkeeping. N. An accounting technique that records each transaction as both a credit and a debit.

double glass. N. Window or door in which two panes of glass are used with a sealed air space between. Synonymous with *insulating glass*.

double-pole reversing switch. See *four-way switch*.

down payment. N. The amount of a home's purchase price a borrower needs to supply up front in cash to get a loan. Most conventional loans require at least 20% of the new home's value as a down payment.

down zoning. N. Rezoning of land from a higher-density use to a lower-density use.

dragnet clause. N. Mortgage clause that compels the mortgagor to pledge additional properties, mortgaged or not, as additional collateral to a different mortgage loan. Failure to pay any of the mortgages can result in a foreclosure on the dragnetted property, even if it is otherwise unmortgaged or its own payments are current.

DRM. ABBRV. Direct reduction mortgage.

dry mortgage. N. Creates a lien against the mortgagor's property but does not permit a lien against his or her personal assets.

drywall. N. A manufactured panel made out of gypsum plaster and encased in a thin cardboard.

DSCR. ABBRV. Debt service coverage ratio.

dual agency. N. Representation of both parties by the same real estate agent or broker.

dual contract. N. Illegal practice of having two contracts for the same transaction. For example, having one contract for a higher amount so that more money can be borrowed.

dual divided agency. N. Representation of two or more parties in a transaction by the same real estate broker.

dual listing. N. Listing where a real estate broker represents both the buyer and seller, creating two principals.

ductile cast iron. See *nodular cast iron*.

due diligence. N. The full investigation of the terms of a contract or a business deal before signing.

due-on-sale provision. N. A provision in a mortgage that allows the lender to demand repayment in full if the borrower sells the property that serves as security for the mortgage.

due-on-transfer provision. N. This terminology is usually used for second mortgages. See also *due-on-sale provision*.

Dutch colonial style

due process. N. The course of legal proceedings established by the legal system of a nation or state to protect individual rights and liberties.

duplex. N. Two separate dwellings under one roof.

duress. N. Act of forcing an individual or business to do something against its will; can be used as a legitimate defense in court to reverse the effect of the compelled act.

Dutch auction. N. The descending-price auction, commonly known in academic literature as the Dutch auction, uses an open format rather than a sealed-bid method. Bidding starts at an extremely high price and is progressively lowered until a buyer claims an item by calling "mine" in some form. When multiple units are auctioned, normally more takers press the button as price declines. In other words, the first winner takes his prize and pays his price and later winners pay less. When the goods are exhausted, the bidding is over.

Dutch colonial style. N. Design that features a barn-like gambrel roof, overhanging eaves, a ground-level front porch and dormers.

Dutch colonial style

early Georgian. N. Type of home prominent in Williamsburg in the 1700s, which was two to three stories with double hung windows and a simple exterior.

Early Georgian

early occupancy. N. Occupation of the property by the buyer before the sale is completed.

earned income. N. Income from personal services as distinguished from income generated by property or other sources. Earned income includes all amounts received as wages, tips, bonuses, other employee compensation, and self-employment income, whether in the form of money, services, or property.

earnest money. See *deposit*.

earning asset. N. An asset that provides income.

earnings. See *income*.

earnings before interest and taxes. See *operating income*. ABBRV. *EBIT*.

earnings before interest, taxes, depreciation and amortization. N. A calculation used to analyze real estate investment trusts. ABBRV. *EBITDA*.

earnings report. N. See *income statement*.

earthquake insurance. N. A policy that provides coverage against damage to a home from an earthquake. Such coverage is usually sold in earthquake-prone areas; homeowners should check to see if their property is located in such an area before buying.

earthquake load. N. Measurement of how much stress a building can stand during an earthquake. Synonymous with *seismic load*.

earthquake strap. N. A metal strap used to secure gas hot water heaters to the framing or foundation of a house, intended to reduce the chances of having the water heater fall over in an earthquake.

easement. N. A right of way giving persons other than the owner access to or over a property.

easement by necessity. N. A legal right to travel to a landlocked parcel of land.

easement by prescription. N. The acquisition of property by adverse land use for a statutory period of time.

easement in gross. N. A personal right to use the land of another but not attached to any one parcel of land.

Eastlake house. N. A nineteenth-century house with plenty of three-dimensional ornamentation, an open front porch, and a turret.

Eastlake house

easy credit. N. Credit that is easy to obtain when lenders reject very few prospective buyers of real estate—this is usually due to an ample money supply and lower interest rates that relax credit standards.

eaves. N. Lower section of the roof forming an overhang and comprised of a fascia, soffit, and soffit molding. The word "eave" comes from the Old English word "off," meaning over.

eave vent. N. Roof opening or opening in an eave that allows for passage of air so that condensation does not form in a tightly insulated house.

EBIT. ABBRV. Earnings before interest and taxes.

EBITDA. ABBRV. Earnings before interest, taxes, depreciation, and amortization.

ECOA. ABBRV. Equal Credit Opportunity Act.

economic base analysis. N. Appraisal method of deriving property values where the current and future economic conditions are measured in a particular area.

economic capacity of land. N. The ability of the size of the property to accommodate the desired economic purpose.

economic force. N. Ability of economic factors to influence the real estate market.

economic indicators. N. Reports made by the government and leading industry groups that measure the past, current, and future direction of the economy and may have an impact on the real estate market. They include the following:

- Measures of general economic performance: gross domestic product, personal income, capital expenditures, corporate earnings, and business inventories.

- Price indexes that illuminate the inflation rate: the Consumer Price Index, a well-known inflation measure for everyday retail pricing, and the Producer Price Index, which monitors raw materials and semi-finished goods at the early stage of the distribution cycle. It reflects changes in the general price level, or the CPI, before they actually occur.

- Measures of labor market conditions: national and state unemployment rates, average manufacturing workweeks, applications for initial jobless claims, and hourly salary rates.

- Money and market indicators: the Dow Jones Industrial Average, the thirty-year Treasury Bill rate, and various measurements of the nation's money supply.

- Combined indicators: the index of leading economic indicators consists of eleven data series comprising the money supply, business formation, stock prices, vendor performance, average work week, new orders, contracts, building permits, inventory change, layoff rate, and change in prices. Business activities are examined as an indication of a change in the economy.

- Measures for major industries: housing starts, resale housing, construction permits, auto and retail sales.

economic life. N. Expected period that property will provide benefits and is typically less than the physical life of the property. Depreciation is usually based on the economic life.

economic surplus. N. The extent to which assets exceed liabilities; profits remaining after subtracting for operating expenses, taxes, interest, and insurance.

economic value. N. The value of an asset, deriving from its ability to generate income.

economies of scale. N. Situation by which the average per square foot cost of construction declines as building size and volume expand.

edge venting. N. Provision of ventilation to attic space by regularly placing vents around the eave line of the roof.

effective age. N. An appraiser's estimate of the physical condition of a building. The actual age of a building may be younger or older than its effective age.

effective debt. N. The total debt that a company owes, including the capitalized value of any lease payments it has to make.

effective gross income. N. Normal annual income including overtime that is regular or guaranteed. The income may be from more than one source. Salary is generally the principal source, but other income may qualify if it is significant and stable.

effective interest rate. N. A consumer-oriented rate that takes into account the projected amount of time the borrower will hold the loan (if he or she holds it to term) as well as the specific costs, fees, and potential rate changes associated with it. The effective rate is not the annual percentage rate (APR). It is similar in that it factors in interest, mortgage insurance, and other fees (including points); however, the APR assumes that a borrower keeps the loan for the entire term, while the effective rate takes into account how long a borrower tells a lender he or she plans to be in the property he or she is financing.

effective net worth. N. Net worth plus subordinated debt.

effective rental income. N. Potential rental income minus vacancy and credit losses.

effective tax rate. N. Tax divided by taxable income equals the effective tax rate. If tax is $30,000 and taxable income is $120,000, then the effective tax rate is 25%.

efficiency ratio. N. Operating expenses divided by fee income plus tax equivalent net interest income.

efficiency unit. N. A small unit—usually without full bath or kitchen facilities—in a multifamily structure.

egress. N. (1) Access from a land parcel to a public road or other means of exit. (2) Right to exit through land owned by another.

ejectment. N. Steps taken to remove someone, who does not have a contractual basis to be there, from the real property.

elbow. N. A plumbing or electrical fitting that allows for changing directions in runs of pipe or conduit.

electrical drawings. N. Plans that show the location of the wiring layout, the types and position of all electrical equipment, and the location of the fixtures.

electrical power. N. Flow of current at a voltage, which is measured in watts (watts = amps x volts).

electrical trim. N. Work performed by the electrical contractor when the house is nearing completion.

electric lateral. N. The trench or area in the yard where the electric service line (from a transformer or pedestal) is located.

electric service. N. Electric power supplied by a utility, which may be any of three capacities, and is either overhead on poles or

electric service panel

buried in the ground. The three capacities are: Two-Wire 115-volt service, where one of the conductors is connected to the assorted electrical devices; Two-Wire 230 volt service, where both wires are connected to the electrical item; and, Three-Wire 115/230-volt service.

electric service panel. N. A panel that translates utility line power into current appropriate for a house, which is then carried through fuses or circuit breakers.

electronic transfer. N. Transfer of a mortgage or other payment automatically by deduction from a checking or savings account.

electrostatic painting. N. In this type of painting, electrically charged powder is sprayed on a surface that is charged with the opposite electrical charge and then bakes on the coating.

eleemosynary. ADJ. Of, relating to, or supported by charity.

elevation. N. Height of a structure above an established reference point.

elevation map. N. Representation on a flat surface of any region that depicts the elevation of that region.

elevations. N. The exterior view of a home design that shows the position of the house relative to the grade of the land.

Elizabethan style. N. English architecture that generally has two levels, with the second level typically overlaying the first story. With a high roof and a sculptured chimney, it usually has half-timber stucco walls.

ell. N. An extension or wing of a house that is perpendicular to the main structure.

ELR. ABBRV. Equivalent level rate.

embankment. N. Mounded soil used as a support along a roadway or to retain water.

emblements. N. Annual crops raised by a land tenant. Even if the lease expires before the crop has matured, the tenant has the right to them.

embrasure. N. Opening for a door or window with the sides slanted so that it is wider on the inside than the outside.

eminent domain. N. The right of a government to take private property for public use upon payment of its fair market value. Eminent domain is the basis for condemnation proceedings.

employer-assisted housing. N. Program that helps employees purchase homes through special plans developed with lenders.

empty nesters. N. Parents who have raised their families and will possibly downsize their dwelling.

encroachment. N. An improvement that intrudes illegally on another's property.

encumbrance. N. Claim, lien, or interest in a property that complicates the title process, interfering with its use or transfer. Anything that affects or limits the fee simple title to a property, such as a mortgage, lease, easement, or restriction.

end loan. N. Conversion of a construction loan to a permanent mortgage on a multi-unit project after all units have been completed.

endorsement. N. (1) A signature on a draft or check by a payee prior to the transfer to a third party. (2) A statement attached to an insurance policy that changes the terms of the policy.

endorser. N. Person who signs over ownership of property to someone else.

endowment. N. Funds or property bestowed upon a person or institution whereupon the income is used to serve a specific purpose for which the endowment was intended.

energy-efficient window. N. Window that has at least two parallel panes of glass so that the loss of heat through the window is slowed.

energy tax credit. N. Tax credit given to encourage the conservation of natural resources as well as the development of alternative resources. See also *business property energy tax credit, residential property energy tax credit.*

enforceable. ADJ. Capable of being enforced; an agreement, debt, judgment, or law that can be put into effect, such as a lien put upon a property upon default of a loan.

engineered 24" framing. N. Building framing using 24" spacing rather than the standard 16" spacing between studs.

engineered masonry. N. Masonry design based on structural analysis.

engineering. N. Science concerned with putting scientific knowledge to practical use. It is divided into different branches: civil, electrical, mechanical, biomedical, or chemical engineering. Activities that can be considered engineering include planning; designing; construction; and management of machinery, roads, bridges, waterways, etc.

engineering controls. N. Valves, switches, regulators, or levers used to manipulate, regulate, or run a system.

entity. N. (1) Separate economic unit subject to financial measurement for accounting purposes. (2) An individual, partnership, corporation, etc., permitted by law to own property and engage in business.

entrance cap. N. A waterproof cap that is placed at the upper part of an electrical mast at the point where the wires are run to the inside electrical meter. Wires hang from the pole to the entrance cap so that the entrance cap is not the low point in the downhill run from the pole because water will run to the low point before dripping to the ground. Wires enter the entrance cap at an upward angle through a tight insulator. Water is further stopped from getting through the entrance cap because of this entrance angle. Synonymous with *weatherhead, mast head, rain cap*.

entrepreneur. N. Individual who starts an enterprise with its associated risks and responsibilities.

environmental codes. N. Laws that govern any part of building that would have an effect on the environment.

environmental impact statement. N. Government-mandated evaluation of the effects a development will have on the environment of a proposed site.

environmentally friendly home construction. N. A construction method that uses environmentally friendly home materials, including recycled items.

Environmental Protection Agency. N. An agency of the U.S. government established to enforce federal pollution abatement laws and to implement various pollution prevention programs. The agency supervises environmental quality and seeks to control the pollution caused by solid wastes, pesticides, toxic substances, noise, and radiation, and has established special programs in air and water pollution, hazardous wastes, and toxic chemicals. It also sponsors research in the technologies of pollution control. Ten regional offices facilitate coordination of pollution control efforts with state and local governments. ABBRV. *EPA*.

environmental site assessment. N. An evaluation of a site, prior to acquisition of title to the property, for the existence of hazardous waste.

EPA. ABBRV. Environmental Protection Agency.

Equal Credit Opportunity Act. N. Federal law requiring creditors to make credit equally available without discrimination based on race, color, religion, national origin, age, sex, marital status, or receipt of income from public assistance programs. ABBRV. *ECOA*.

equalization. N. Mass appraisal of all property within a jurisdiction for the purpose of equalizing values to assure that each taxpayer is bearing a fair share of the tax load.

Equifax. N. Equifax Credit Information Services, Inc.; one of the "Big Three" credit reporting bureaus that operate nationwide.

equitable conversion. N. Legal doctrine applied in some states in which, under a contract of sale, buyer and seller are treated as though the closing had taken place, in that the seller in possession has an obligation to take care of the property. Even though the legal title has not passed, the law holds that there has been an equitable conversion that effectively vests the property in the buyer.

equitable lien. N. Written contract or court judgment placing a lien on a parcel of property as collateral for a loan.

equitable owner. N. The person identified to receive the benefit of property held in trust.

equitable title. N. The right to demand that title be conveyed upon payment of the purchase price.

equity. N. The difference between the fair market value of the property and what the borrower still owes on the mortgage. A lender determines how much equity is in the home by taking the appraised value of the home and subtracting any mortgage debt. For example, if a house is valued at $150,000 and the mortgage balance is $90,000, the borrower has $60,000 in equity. That equity is considered the basis for any second mortgages or home equity lines of credit the borrower may ask for later.

equity buildup. N. The increase in a person's equity in real estate due to the reduction in the mortgage loan balance and price appreciation.

equity cushion. N. Ownership interest in property that is above the minimum need to meet uncertainties or a downward trend in a real estate market.

equity in property. N. Amount by which the appraised value of property exceeds the debt balance. If property has a fair market value of $500,000 while the mortgage balance is $200,000, the owner's equity in the property is $300,000.

equity kicker. N. Privilege of a real estate investor or lender to participate in the profitability generated from property, in addition to any principal, interest, or dividends. Synonymous with *kicker*, *participation loan*.

equity lending. N. Bank financing to a homeowner based on his or her dollar equity in the home. If a home is worth $500,000 and the owner owes $200,000, his or her equity is $300,000.

equity loan. N. A loan to a home or condominium owner that is secured by the lender against the equity the owner has built up in the property.

equity of redemption. N. Borrower's right to redeem his or her property by immediately paying off the loan balance and any related costs.

equity participation. N. Lender has an equity interest in the property that is the subject of the loan. This is in addition to principal and interest payments on the mortgage. The lender shares in the increase in market price of the property as well as any net income generated.

equity purchaser. N. An individual or business that buys someone else's equity in property but may not assume any responsibility for a loan balance.

equity rate of return. N. Return before taxes on the capital invested in real estate property.

equity REIT. N. Type of real estate investment trust (REIT) whose investment money is used for the purchase of a portfolio of specific properties to be managed in order to generate investment return through current income and capital gain. Synonymous with *equity trust*.

equity sharing. N. Arrangement whereby a party providing financing gets a portion of the ownership.

equity-to-value ratio. N. Ratio of the purchase price of property to its total appraised value. If property is appraised at $500,000 and the price paid is $400,000, the ratio is 80%.

equity trust. N. See *equity REIT*.

equity yield rate. N. The rate of return on the equity portion of an investment, taking into account periodic cash flow and the proceeds from resale. Considers the timing and amounts of cash flow after annual debt service, but not income taxes.

equivalent level rate. N. Flat rate per square foot that will equal the same total present value as a proposed lease's variable cash flows. ABBRV. *ELR*.

errors and omissions insurance. N. A policy that pays for any mistakes a builder or architect makes in a project. Anyone hiring a builder or an architect should ask if the professional has this coverage before hiring him or her.

escalation clause. N. (1) Provision in a lease that requires the tenant to pay more rent based on an increase in costs. (2) A provision in a loan agreement or mortgage in which the entire debt becomes immediately due upon the occurrence of an item, such as missing three consecutive monthly payments or when the current ratio falls below 1.0.

escape clause. N. Provision in a contract that allows one or more of the parties to cancel all or part of the contract if certain events or situations do or do not happen.

escheat. N. When the ownership of a property reverts to the state because the owner dies without leaving a will.

escrow. N. The holding of documents and money by a neutral party for a real estate transaction. This ensures that all conditions of the sale are met.

escrow account. N. An account that the lender or mortgage service establishes to hold funds for the payments of property taxes and insurance. Synonymous with *impound account*.

escrow agent. N. A neutral third party who ensures that all conditions of a real estate transaction are completed satisfactorily.

escrow analysis. N. The periodic examination by the lender of an escrow account for purposes of determining if the amount withheld from a borrower's monthly mortgage payment is sufficient to pay for expenses such as property taxes and insurance.

escrow closing. N. When all the conditions of a real estate transaction are completed and title of the property is transferred to the buyer, escrow is considered closed.

escrow company. N. A company that acts as a neutral third party, ensuring that all conditions of a real estate transaction established by the buyer, seller, and lender are fulfilled.

escrow fees. N. Amount earned by the escrow agent for accumulating and monitoring data from various sources and distributing it to the parties.

escrow payment. N. Funds withdrawn from a borrower's escrow account by the mortgage service to pay property taxes and insurance.

escrow statement. N. Declaration by an escrow agent that instruments or property are being held in accordance with the agreement to the parties in a real estate deal.

estate. N. A taxable entity that is established upon the death of a taxpayer. It consists of all the decedent's property and personal effects. The estate exists until the final distribution of its assets to the heirs and other beneficiaries. The executor must complete his or her administration of the estate with the disbursement of assets and the filing of a tax return.

estate at sufferance. N. The wrongful occupancy of property by a tenant after the lease has expired.

estate at will. N. The occupation of real estate by a tenant for an indefinite period, terminable by one or both parties at will.

estate for life. N. Interest in property that terminates upon the death of a specified person.

estate for years. N. An interest in land allowing possession for a specified and limited time.

estate in revision. N. An estate left by the grantor for him- or herself, to begin after the termination of some particular estate granted by him or her. For example, a landlord has an estate in revision, which becomes his or hers to possess when the lease expires.

estate of inheritance. N. An estate that descends to heirs in perpetuity.

estate on condition. N. Land property estate contingent upon the occurrence or lack of occurrence of a particular event whereupon it can be created, augmented, or dismantled.

estimate. N. An appraisal value of property, an approximation of market values. Alternately, to calculate the approximate computation of the cost of completion of construction.

estimated closing costs. N. An estimate of the expenses incidental to the sale of real estate, including loan, title, and appraisal fees. These costs exist in addition to the price of the property and are paid at closing. Some are one-time expenses and some are recurring.

estimated cross costs of buying. N. An estimate of the total principal and interest payments over the number of years that a homeowner plans to own his or her home.

estimated hazard insurance. N. An estimate of hazard insurance, known as homeowner's insurance or fire insurance, needed to cover physical damage such as from fire and wind. Coverage is usually required to equal the replacement value of the home.

estimated increase in equity. N. A specified property value increased by a selected rate of appreciation for a specific number of years.

estimated net costs of buying. N. The estimated gross costs of buying minus estimated tax savings and the estimated increase in equity.

estimated property taxes. N. An estimate of property taxes to be paid. The amount is based on local tax rates and assessed property value (based on most recent sale price plus assessment updates).

estimated taxes and insurance. N. Calculation, used by a lender, of estimated taxes and insurance, which is used to evaluate a borrower's effective monthly housing expense.

estimated tax savings. N. The amount of tax a renter would save instead of owning a home, based on property taxes and interest paid.

estimated total costs of renting. N. The total current rental payments for the same number of years a homeowner would plan to own a home increased by a yearly rental increase adjustment.

estimated total savings. N. The estimated net costs of renting minus the lower net cost of buying.

estimated useful life. N. The period of time over which an asset will be used by a particular taxpayer. Although that period cannot be longer than the estimated physical life of an asset, it can be shorter if the taxpayer does not intend to keep the asset until it wears out. The estimated useful life of an asset is essential to determining the annual tax deduction for depreciation and amortization.

estoppel. N. (1) Prevention of a person from making a statement of affirmation or denial because it is contrary to a previous statement. (2) The barring of an act.

estoppel by deed. N. Restraining a person or business from denying an appropriate conveyance of property

estoppel certificate. N. Mortgagor's signed statement that the stated remaining balance of a mortgage is correct and it is a property lien, which later prevents him or her from stating that the facts were misrepresented, therefore making the mortgage invalid.

estover. N. Right of a tenant to make use of a property's wood- or food-producing capacity to provide for his or her own necessities.

et al. ABBRV. (*Latin*) And others; used to indicate the presence of more people than are named.

et con. ABBRV. With husband.

et ux. ABBRV. With wife.

evict. V. To remove a tenant through a legal process.

eviction. N. Legal procedure for removal of a tenant for reasons that would include, but not be limited to, failure to pay rent. See also *actual eviction, constructive eviction, partial eviction.* Synonymous with *summary possession.*

evidence of title. N. A document, such as a deed, that demonstrates property ownership.

examination of title. N. (1) Title company review of public records and other documents to determine the chain of ownership of a property. (2) The report on the title of a property from the public records or an abstract of the title.

excavation. N. The process of clearing trees, removing topsoil, and grading land before the foundation is laid.

exception. N. (1) Waiver of a requirement in an agreement. (2) A right or portion of property reserved to the grantor in a conveyance by deed.

exceptional depreciation. N. Damages to a building that exceed that of normal wear and tear.

excess condemnation. N. Taking more property in a condemnation proceeding than was originally required or planned.

excess depreciation. N. Costs taken over and above what one is entitled to and that can occur either by claiming depreciation costs exceeding the actual depreciable value or by depreciating items that cannot be depreciated.

excess income. Rental income received from property that exceeds the costs of owning and maintaining the property.

exchange. N. A transfer of property for other property or services. Exchanges of like-kind property are a popular method for deferring taxes.

excluded gain. N. Generally applies to gains realized on the sale of a principal residence.

exclusionary zoning. N. Property zoning that has the net effect, intended or not, of excluding the poor and minority groups from living in a particular area. Large property size requirements are often responsible for exclusionary zoning.

exclusive agency. N. Employment of a particular broker. If a sale by another broker is accepted, both are entitled to commissions.

exclusive buyer's agent. N. An agent or company that works exclusively for buyers. They do not represent sellers or list properties. ABBRV. *EBA*.

exclusive listing. N. A written contract that gives a licensed real estate agent the exclusive right to sell a property for a specified time, but reserves the owner's right to sell the property alone without the payment of a commission.

exclusive right to sell. N. Contract with a real estate agent that pays a commission to the agent even if the property is sold to a buyer found by the owner.

exclusivity. N. A restriction on the tenant's right to open a competing retail outlet within a certain area of leased premises.

exculpatory clause. N. A provision in a mortgage that allows the borrower to surrender the property to the lender without personal liability for the loan.

execute. V. To sign, complete, perform, and carry out all the terms of a contract, including signing the contract and delivering it to the proper party.

executed contract. N. A contract whose terms and provisions have been completely fulfilled and satisfied by all involved parties.

executor. N. Someone appointed to carry out instructions left in a will. If no executor is named in the will, one is named by probate court.

executor's deed. N. A deed to convey property, which is done by the executor of a will once it is authorized by the probate court.

executory cost. N. Cost excluded from the minimum lease payments to be made by the lessee in a capital lease. The lessee reimburses the lessor for the lessor's expense payments.

exemplary damages. See *punitive damages*.

exempt. ADJ. Describes real estate that is not subject to property tax such as that owned by nonprofit entities, including charitable, governmental, and religious institutions.

exemption. N. (1) An amount provided by law that reduces taxable income or taxable value. (2) Removal of property from the tax base, partially or completely.

exhaust fan. N. Ventilating device that removes water vapor, odor, and smoke.

expansible house. N. House designed to be easily expanded, such as a basement that can be finished or an attic that can be expanded into more bedrooms.

expansion option. N. Right granted by the landlord in the lease that provides the option of adding more space to the premises.

expenditure. N. A payment, or the promise of a future payment.

expense. N. Any operating cost, such as rent, utilities, and payroll, as distinguished from capital expenditure for long-term property and equipment. For mutual funds, a fund's cost of doing business. All of a mutual fund's expenses are disclosed in its prospectus as a percentage of assets.

expense ratio. N. A comparison of operating expenses to potential gross income. This ratio can be compared over time and with that of other properties to determine the relative operating efficiency of the property considered.

expenses. N. For federal income tax purposes, expenses are divided into four categories: trade or business expenses; expenses in connection with production of income, in connection with management, conservation, or maintenance of property held for production of income; expenses in connection with the determination, collection, or refund of any tax; and, personal, family, or living expenses. Expenses in the first three categories are generally deductible in determining taxable income, while expenses in the fourth category are not deductible, except in a few cases (e.g., medical expenses, charitable contributions, etc.) in which they are specifically allowed by law. Expenses are to be distinguished from capital expenditures.

expenses of sale. N. When paid by the seller, these expenses reduce the sale price of property. Examples are commissions to a broker or real estate agent, title search, title insurance, legal fees, and transfer taxes.

expense stop. N. Fixed amount in a lease that serves as a limit on the amount for which the tenant is responsible, in the event expenses exceed a prestated amount.

expensing. See *Section 179 Expense Deduction*.

Experian. N. Formerly known as TRW Information Systems and Services, Experian is one of the "Big Three" credit reporting firms.

ex post facto. N. (*Latin*) An act occurring after the fact.

exposure. N. (1) Person or business susceptible to loss on an investment, such as a high-risk speculation. (2) The advertising, whether free or paid, of property that is for sale. (3) The condition of being open to the elements; unhidden from view.

express agreement or contract. N. Contract where all parties express their intentions in words orally or in writing.

extended coverage. N. Protection over and above that of a standard policy or warranty and that covers things that might otherwise be excluded.

extension option. N. Agreed continuation of occupancy under the same conditions as outlined in previous agreements.

extension, tax. N. A filing made with the Internal Revenue Service to pay taxes in full after the deadline. In the extension form, a taxpayer must make a reasonable estimate of his or her tax liability for the year. Failing to do so will result in a penalty. Anyone who has already paid 90% of the estimated tax through withholding will not be charged a penalty.

extraordinary item. N. A nonrecurring event that materially affects a company's finances in a reporting period. Must be explained in the annual report or quarterly report.

extras. N. Additional work requested of a contractor, not included in the original plan, that will be billed separately and will not alter the original contract amount, but will increase the cost of building the home.

façade. N. External front of a building that faces the street or courtyard and is usually used to describe large, elegant buildings. Façade materials include wood, brick, glass, masonry, aluminum, etc.

face brick. N. Exterior decorative surface, which is made of brick that is not rendered, painted, or plastered, and is made of various brick materials, including clay, to give a desired effect.

face concrete. N. The finish of the front and all vertical sides of a concrete porch, steps, or patio.

faced masonry. N. Masonry structure that has different types of material as backing and facing bonded together, such as brick on concrete.

face velocity. N. Measurement of the air velocity as measured at the face of the inlet or outlet in a heating, ventilation, or air conditioning system.

facilitator. N. Real estate professional who aids in a transaction but does not have an agency relationship with that party and can be known as an intermediary or transaction broker.

factoring. N. (1) The purchase of the accounts receivable of a business. (2) Taking the accounts receivable of a business as collateral for a loan.

factor of safety. N. The ratio of the maximum strength of a piece of material or a part to the probable maximum load to be applied to it. If a maximum of two thousand pounds can be tolerated, a load of five hundred pounds will have a four to one factor of safety.

Fair Credit Billing Act. N. Federal law governing credit and charge card billing errors. If the credit card company violates this law, consumers can sue for damages.

Fair Credit Reporting Act. N. A consumer protection law that regulates the disclosure of consumer credit reports by consumer-credit reporting agencies and establishes procedures for correcting mistakes on one's credit record.

Fair Debt Collection Practices Act. N. Federal law that outlaws debtor harassment and regulates collections agencies, original creditors' collection offices (if separate), and creditors' lawyers. Original creditors may be covered under state law.

fair housing. N. Housing policy where purchase or lease is available to everyone regardless of race, national origin, color, or religion and affirmative action is actively pursued. Synonymous with *open housing, open occupancy*.

Fair Housing Act. N. Federal law making it illegal to refuse to rent or sell to anyone based on race, color, religion, sex, or national origin. Amendments made in 1988 expanded protections to include family status and disability.

Fair Isaac Corporation. N. Founded in 1956, Fair Isaac is a leading company that creates the credit score technology used by lenders. The company's predictive modeling, decision analysis, intelligence management, decision management systems, and consulting services are behind more than twenty-five billion lending decisions a year.

fair market rent. N. The rent a property commands in a free and open market setting.

fair market value. N. The amount at which property would change hands between a willing buyer and a willing seller, neither being under compulsion to buy or sell and both having reasonable knowledge of the relevant facts.

fair rental value. N. The amount the owner of property could reasonably expect to receive from a stranger for the same type of lodging; generally, the amount at which a home with its furnishings could be rented to a similar size family in a similar location.

falling out of escrow. N. Situation where one of the parties is unable to satisfy the conditions of the purchase and sale contract.

false personation. N. A criminal act of falsely representing another individual to gain a profit or an advantage.

family limited partnership. N. A limited partnership whose interests are owned by members of the same family so that gift and estate taxes may be reduced, although the freedom or transferability of ownership is not available.

Fannie Mae. ABBRV. Federal National Mortgage Association.

FAR. ABBRV. Floor area ratio.

FARM Credit Agency. N. An agency of the federal government that makes mortgage loans on rural property to farmers and to individuals who provide services to farmers and ranchers. Loans are made at below-market interest rates. Borrowers are required to purchase stock in their local land bank association, which serves as additional security for the loan.

Farmer's Home Administration. N. U.S. Department of Agriculture agency, which provides credit to qualified farmers and rural residents. ABBRV. *FMHA*.

FASB. ABBRV. Financial Accounting Standards Board.

fascia. N. Horizontal boards attached to rafter or truss ends at the eaves and along gables. Roof drain gutters are attached to the fascia.

FDIC. ABBRV. Federal Deposit Insurance Corporation.

feasibility study. N. A determination of the likelihood that a proposed development will fulfill the objectives of a particular investor. This study should estimate the demand for the product, the absorption rate, legal considerations, cash flow, and approximate investment returns likely to be produced. Analysis is also made of alternative means of accomplishing the task.

federal agency securities. N. Debt instruments of U.S. agencies such as the Federal Home Loan Bank, the Federal National Mortgage Association, the Federal FARM Credit Bureau, and the Tennessee Valley Authority. Although these issues are not direct obligations of the U.S. Treasury, they still have a high credit rating.

Federal Deposit Insurance Corporation. N. A public corporation established in 1933 that insures up to $100,000 for each depositor in most commercial banks and savings and loan associations. It has its own reserves and can borrow from the U.S. Treasury. ABBRV. *FDIC*.

Federal Home Loan Bank Board N. Federal agency that monitors the federal savings and loan associations and federally insured state-chartered savings and loan associations and acts as a central bank. In addition, it operated the Federal Savings and Loan Insurance Corporation. ABBRV. *FHLBB*.

Federal Home Loan Bank System. N. A federally created banking system intended to assure liquidity to qualified savings and loan lenders; twelve regional Federal Home Loan Banks have been directed by the Federal Home Loan Bank Board since 1932.

Federal Home Loan Mortgage Corporation. N. An agency that buys loans that are underwritten to its specific guidelines, pools them, and sells shares to investors. These guidelines are an industry standard for residential conventional lending. ABBRV. *FHLMC, Freddie Mac.*

Federal Housing Administration. N. A federal agency within the Department of Housing and Urban Development (HUD) that insures residential mortgage loans made by private lenders and sets standards for underwriting mortgage loans. ABBRV. *FHA.*

Federal Housing Administration Insured Mortgages. N. Mortgages insured by the Federal Housing Administration (FHA). FHA insurance is intended to make more housing available and to safeguard the lender against risk of nonpayment.

Federal Housing Finance Board. N. Federal agency created under the Financial Institutions Reform, Recovery and Enforcement Act to regulate and supervise the twelve district Federal Home Loan Banks. The functions of the Federal Housing Finance Board were the responsibility of the Federal Home Loan Board prior to passage of the act.

federally related mortgage. N. A mortgage loan that is, in some way, subject to federal law because it is guaranteed, insured, or otherwise regulated by a government agency.

federally related transaction. N. Real estate transaction that is overseen by a federal agency.

Federal National Mortgage Association. N. A congressionally chartered, shareholder-owned company buys mortgages that are underwritten to its specific guidelines from lenders and resells them as securities on the secondary mortgage market. These guidelines are an industry standard for residential conventional lending. ABBRV. *FNMA, Fannie Mae.*

Federal Reserve Board. N. Group of economists who set the nation's monetary policy through its ability to control interest rates, thereby controlling inflation.

Federal Savings and Loan Association. N. Charter issued by the Office of Thrift Supervision, under the U.S. Department of Treasury, to an institution to act as a savings and loan association. A federally chartered savings and loan association, in contrast to one with a state charter, may have the ability to branch across state lines as well as make certain investments that a state-chartered thrift institution cannot.

federal style. N. American architectural style, which evolved after the Revolutionary War and includes bigger windows and a glass-surrounded front doorway, topped with an arched window.

federal tax lien. N. A lien placed on an individual's real property by the federal government for federal income or estate tax violations. If these taxes are not paid, the government may seek a tax warrant causing a federal tax lien to be placed against the taxpayer's property. In the event of death, the estate is liable for the lien.

Federal Trade Commission. N. Government agency that regulates companies and industries, including collections agencies, time-share operators, etc. ABBRV. *FTC*.

fee ownership. N. Form of ownership that gives the owner complete control, including over the development of an inheritable estate, also known as a fee estate.

fee simple. N. This type of ownership, which is the maximum interest a person can have in a piece of real estate, entitles the owner to use the property in any manner as long as it is in accordance with state and local laws.

fee simple absolute. N. An estate limited absolutely to an owner and his or her heirs in perpetuity and without limitation. This status entitles the owner to full ownership of the property and the unrestricted ability to divide it among his or her heirs.

fee simple conditional. N. A fee estate conditioned by the provisions of the grantor or the grantor's heirs that some action occur in order to complete its conveyance. If this condition does not occur, the estate returns to the original grantor.

fee simple defeasible. N. Fee simple ownership that can be defeated and returned to the grantor should a particular event occur.

fee simple determinable. N. A fee estate limited by the happening of a certain event.

feng shui. N. Ancient Chinese philosophy that the positioning and physical characteristics of a home affect the fortunes and well-being of the owner.

FHA. ABBRV. Federal Housing Administration.

FHA loans. N. Mortgage loans that are insured by the Federal Housing Administration (FHA). The FHA operates loan plans for purchasers of rural property as well as provides low-rate mortgages to buyers who make down payments as low as 3%. With FHA insurance, a borrower can purchase a home with a down payment from 3% to 5% of the FHA-appraised value or the purchase price, whichever is lower. FHA mortgages have a maximum loan limit that varies depending on the average cost of housing in a given region. In general, the loan limit is less than what is available with a conventional mortgage through a lender.

FHLBB. ABBRV. Federal Home Loan Bank Board.

fidelity. N. (1) Accuracy of a description, translation, or reproduction. (2) Faithful devotion to duty or obligation.

fidelity bond. N. Insurance coverage purchased by an employer to cover employees who are entrusted with valuable property or funds, to protect against specified losses arising from any dishonest act by these employees.

fiduciary. N. Person or institution acting in a legal capacity, in the best interests of someone including the holding or administration of property owned by another.

fiduciary duty. N. The holding in trust of something by one person for another. Also applies to legal, real estate, and business relationships.

fifteen-year mortgage. N. A fixed-rate, level-payment mortgage loan where, for a slight increase in monthly payments, the loan can be paid off in only fifteen years. The overall savings in interest paid to the lender between the fifteen-year and thirty-year mortgage can be quite substantial without making the monthly payment significantly higher.

final value estimate. N. Final property appraisal estimate arrived at with the use of appropriate appraisal methods.

final walk-through inspection. N. A sales contract should include a clause that allows a buyer to examine the property he or she wants to purchase within the twenty-four hours before closing. Real estate professionals typically accompany their clients on this walk-through, which is the last chance to ensure that the seller has vacated the house and left behind whatever property was agreed upon before papers are signed. During the walk-through, a buyer should make sure to check that all lights, appliances, and plumbing fixtures are in working order and double-check that all conditions of the sales contract have been met. Plan to delay closing if problems are discovered and not cleared up before a specific time.

finance charge. N. Interest and any other charges, including points that make up the fees incurred when borrowing money.

financed closing costs. N. Costs for closing of title that are added to the loan amount rather than being paid up front. This practice adds to the amount borrowed, increasing the monthly payment.

Financial Accounting Standards Board. N. Independent agency that establishes generally accepted accounting principles. ABBRV. *FASB*.

financial asset. N. A nonphysical asset, such as a security, certificate, or bank balance; the opposite of a nonfinancial asset.

financial calculator. N. Calculator that has numerous built-in financial functions including cash flow analysis, mortgage amortization, present and future yield, yield to maturity, and many other business statistics and financial ratios.

financial capital. N. Funds that are available to acquire real capital.

financial condition. N. The status of a firm's assets, liabilities, and equity positions at a specific point in time, often described in a financial statement.

financial feasibility. N. The ability of a proposed land use or change of land use to justify itself from an economic point of view.

financial index. N. An index is a number to which the interest rate on an adjustable rate mortgage (ARM) is tied. It is generally a published number expressed as a percentage, such as the average interest rate or yield on U.S. Treasury bills. A margin is added to the index to determine the interest rate that will be charged on ARMs. This interest rate is subject to any caps associated with the mortgage. The interest rate changes on an ARM are tied to some type of financial index. When comparing ARMs, look at how the index to which it is tied has performed recently. A lender can provide information on how to track the index and a history of the index it uses. Some of the most common types of indexed ARMs include Treasury-Indexed ARMs, CD-Indexed ARMs, Cost of Funds-Indexed ARMs, and LIBOR-Based ARMs.

Financial Institutions Reform, Recovery and Enforcement Act. N. A federal law passed in 1990 that restructured the regulatory and deposit insurance apparatus dealing with savings and loan associations (S&Ls) and changed the rules under which federally regulated S&Ls operate. ABBRV. *FIRREA*.

financial intermediary. N. A firm, such as a bank or savings and loan association, that performs the function of collecting deposits from individuals and investing them in loans and other securities.

financial leverage. N. The use of borrowed money to complete an investment purchase. Synonymous with *trading on equity*.

financial markets. N. Institutions acting as intermediaries between suppliers and users of money. They are wholesale and retailers of funds. The financial markets are where those wanting funds are matched with those having surplus funds.

financial statement. N. Report that shows income and expenses for an accounting period and normally consists of a balance sheet, income statement, and statement of cash flows. A bank may request a financial statement from a prospective borrower for commercial property or any other business use.

financial structure. N. The right side of a firm's balance sheet, detailing how its assets are financed, including debt and equity issues.

financing. N. The loaning and borrowing of money to buy an item.

finder's fee. N. A fee or commission paid to a mortgage broker for finding a mortgage loan for a prospective borrower.

fire brick. N. Brick made of refractory ceramic material that will resist high temperatures. Used in fireplaces and boilers.

fire division wall. N. A wall designed and rated to delay passage of fire, which extends continuously from the bottom to the top of a structure.

fire door. N. A door designed to resist the passage of fire. Fire doors are rated by the amount of time they can resist the penetration of fire with the time ranging from one-half to three hours. Fire doors are used to close openings in firewalls, so that the door area is no more vulnerable to fire than the wall.

fire extinguisher. N. Device containing fire-suppressing material under pressure, which is directed at the fire to extinguish it by means of oxygen deprivation and/or cooling. Extinguishers are rated as Grade A if they are to be used on ordinary material fires; Grade B if they are to be used on fluid fires; Grade C if they are to be used on electrical fires; or, Grade D if they are to be used on materials that need an extinguishing compound to absorb heat and do not react with the fuel.

fire line. N. A wet standpipe, i.e., a vertical pipe that is always full of water, which reaches to the upper floors of a building and can be immediately accessed to distribute water in the event of a fire.

fire rating. N. Rating system that shows the fire resistance of a material or system as tested by a recognized laboratory against applicable American Society for Testing and Materials standards.

fire stop. N. A solid, tight closure of a concealed space, placed to prevent the spread of fire and smoke through such a space.

firm commitment. N. A written promise, made by the lender, to loan money. It usually contains all of the terms relevant to the transaction.

first-generation space. N. New space in a building that has never been occupied by a tenant.

first mortgage. N. Primary mortgage on a property, which takes priority over any other liens and is satisfied before any secondary liens. In the case of a foreclosure, the first mortgage will be repaid before any other mortgages.

first refusal right. N. Being offered the right to buy something before it is offered to others; the opportunity of a party to match the terms of a proposed contract before the contract is executed.

first user. N. Initial user of real estate, such as the first occupant of a newly built home.

fiscal. ADJ. Pertaining to money, especially government taxation and spending policies.

fiscal year. N. An accounting period of 365 days (366 in leap years), but not necessarily starting on January 1.

five Cs of credit. N. A historic guideline lenders have used to award credit that still holds today. The five Cs are character (willingness to pay); capacity (financial cash flow); capital (wealth); collateral (security); and, conditions (economic status).

fixed asset. N. A long-term, tangible asset held for business use and not expected to be converted to cash in the current or upcoming fiscal year, such as real estate, business equipment, and furniture. Synonymous with *long-term asset*.

fixed-charge coverage ratio. N. Profits before income taxes and interest payments, divided by long-term interest, for a given period of time.

fixed cost. N. A cost that does not vary depending on production or sales levels, such as rent, property tax, insurance, or interest expense.

fixed expenses. N. Those expenses that remain the same regardless of circumstances.

fixed installment. N. Monthly home loan payments.

fixed-payment mortgage. N. A loan secured by real estate, which features a periodic payment of interest and principal that is constant over the term of the loan. All fixed-payment mortgages are fixed-rate mortgages, but some fixed-rate mortgages may have variable payments, such as a graduated payment mortgage.

fixed-period adjustable rate mortgage. N. This type of adjustable rate mortgage maintains the same initial interest rate for the first three, five, seven, or ten years of the loan, depending on the term chosen. Afterward, the rate adjusts annually, and can move up or down as market conditions change—borrowers should make sure there is a cap that prevents such a mortgage from becoming unaffordable.

fixed-rate mortgage. N. Loan with an interest rate that remains at a specific rate for the entire loan. Approximately 75% of home mortgages are fixed-rate.

fixer-upper. N. A home that needs repair and remodeling and sells at a below-market price.

flame spread rating. N. Tests, done in accordance with ASTM Standard E 84, for establishment of fire-resistant values of building materials by measuring how fast and far flames will spread over certain surfaces.

flapper ball. N. Rubber valve covering the flush valve in a toilet tank.

flashing. N. The prevention of water seepage by installing metal strips around chimneys, vents, windows, doors, and skylights, and along seams in the roof and beneath shingles. Flashing also provides a drainage passageway between joints, most commonly the joint between a roof and a wall.

flat. N. An apartment within a multifamily house, usually on one floor.

flat fee. N. Set amount charged by a broker.

flat lease. N. Lease agreement that has level payments during the contractual period and does not have an escalation clause, which would allow for increased costs due to increases in inflation, taxes, or other related costs. Synonymous with *straight lease*.

flexible loan insurance program. N. A graduated payment mortgage (GMP) developed to overcome the negative amortization aspects of the GMP. The buyer's own down payment is deposited in a pledged, interest-bearing account, where it is used as both cash collateral and a source of supplemental payments during the initial years of the loan. During this time, predetermined amounts are withdrawn by the lender from the savings account and added to the borrower's reduced payment, making a full mortgage payment. Decreasing every month, it disappears at the end of a predetermined period. Using this type of program is likely to make a borrower able to qualify for a larger loan than with a conventional fully amortized mortgage. ABBRV. *FLIP*.

flexible payment mortgage. N. Loan allowing the borrower to pay only the interest for the first few years of the loan.

flexible rate mortgage. N. Mortgage with an interest rate that changes based on certain events, such as changes in the prime rate. ABBRV. *FRM*.

flex space. N. A structure that allows its occupants the flexibility of using its space for a variety of uses, usually commercial. Such buildings allow tenants the ability to use space for offices, showrooms, manufacturing, laboratory, warehouse distribution and gallery space. These structures are generally built with little or no common areas, load-bearing floors, loading dock facilities, and high ceilings.

floating interest rate. N. An interest rate that is not fixed over the term of a loan, bond, or other fixed-income security, but is allowed to vary according to the change in a specified index, such as the prime interest rate or the Treasury bill rate.

flood certification. N. The determination as to whether or not a property is located in a flood zone. If it is, the lender will require federally provided flood insurance.

flood insurance. N. Special coverage that is required for property in a designated flood plain or zone since that risk is typically not covered under standard hazard insurance.

flood plain or zone. N. Level land area subject to periodic flooding from a contiguous body of water. During the flood stage, the property may be underwater.

floor area ratio. N. The ratio between a structure's total floor area and the total land area of the land upon which it is constructed. The FAR is calculated by dividing the total building floor area by the total building lot square area: floor area ratio = building floor area ÷ building lot area. A maximum allowable floor area ratio is typically specified by the local building code or zoning. ABBRV. *FAR.*

floor joists. N. Beams that provide structural floor support; the flooring is directly attached to the floor joists.

floor loan. N. The minimum amount of money a lender is willing to provide on a commercial loan for a building that is to be tenant-occupied. This loan is progressively funded as the building is constructed and occupied.

floor plan. N. The arrangement of rooms in a structure. A two-dimensional scale drawing of the arrangements, size, and orientation of doors, rooms, walls, and windows of a single floor of a building structure.

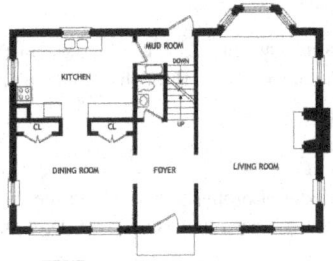

Floor Plans

Florida rooms. N. Glass-enclosed porches at the rear or side of a home, usually in warm weather areas.

flue. N. Large pipe through which fumes escape from a gas water heater, furnace, or fireplace.

flue damper. N. An automatic door located in the flue that closes it off when the burner turns off; purpose is to reduce heat loss up the flue from the still-warm furnace or boiler.

flue lining. N. Heat-resistant lining, usually of fire clay or terra cotta pipe, used for the inner lining of chimneys.

flue pipe. N. Airtight conduit constructed within a chimney using fireproof materials to carry away combustion gases and smoke occurring in a furnace or firebox.

flush door. N. Door with plywood facing over the internal core of wood or wood products. A hollow core door is one with plywood facing over framework without a solid core.

FNMA. ABBRV. Federal National Mortgage Association.

forbearance. N. A lender's decision to refrain from taking legal action on a delinquent loan in hopes the default will be remedied in a timely manner.

forced air heating system. N. A heating system that circulates warm air, from a heat source, through the ducting by means of a blower fan.

forced hot air. See *forced air heating system*.

forced sale. N. Sale of property where the seller is under duress and current market prices and conditions determine the selling price.

force majeure. N. An unavoidable cause of delay or of failure to perform an obligation in time due to an unpreventable, overwhelming, and irresistible force. Synonymous with *act of God delay*.

foreclosure. N. Legal process in which a lender ends the borrower's interest in a property after a loan is defaulted. The lender takes the property and then sells it to cover the mortgage amount and legal costs from the proceeds of the sale. Any other proceeds may be returned to the borrower. Foreclosure is usually a last step lenders will take after efforts to work out a delinquent loan have been exhausted.

foreclosure sale. N. The public sale of a property after the legal process, by which the lender ends the borrower's interest in a property, is completed.

forfeit. V. To relinquish a right.

forfeiture. N. The loss of money, property, rights, or privileges due to a breach of legal obligation.

formaldehyde. N. Colorless gas chemical that was used in foam insulation for homes until the early 1980s and is emitted by some construction materials. It is suspected of causing cancer, and it can also irritate the throat, nose, and eyes. A qualified inspector can determine if the gas is present in the home. See also *contingency*.

for sale by owner. N. Homeowner acting as salesperson for his or her own property. No listing commission is paid but a sales commission may be paid to a buyer's agent. ABBRV. *FSBO*.

foundation. N. (1) The support structure of a house. (2) The base or portion of a structure that is in contact with the ground, usually extending below grade. (3) A support on which something stands.

foundation bolt. N. A bolt that is set into wet concrete to be used for the attachment of the pressure-treated boards (called the mud sill or sill plate) once the concrete has hardened. Once in place, these boards will be the base of the framing structure.

foundation plan. N. A drawing used for construction, which shows all the dimensions and placement of the foundation.

four-way switch. N. Switching device that uses three switches to permit an outlet to be operated from all three switches. This electrical switch is used in conjunction with two three-way switches in cases where three points for controlling an electrical circuit are needed. Synonymous with *double-pole reversing switch*.

frame. N. (1) The basic, skeletal structure around which a building is built and which gives it its shape or form. (2) The border or case into which a window or door is set that serves as a structural support. V. To enclose in or to provide a border for something.

Freddie Mac. ABBRV. Federal Home Loan Mortgage Corporation.

free and clear title. N. Property title that has no encumbrances, including mortgages, judgments, and financial liens.

free cash flow. N. Operating cash flow (net income plus amortization and depreciation) minus capital expenditures and dividends.

freehold estate. N. Estate in which ownership is for an indeterminate length of time; unlimited interest in a property. Freehold estates include freehold in deed, a fee simple estate; freehold in law, an inheritable estate; and, determinable freeholds, a life estate.

French provincial. N. A formal, two-story house with a high, steep roof and curve-headed upper windows that come through the cornice.

frieze. N. In home construction, a horizontal band connecting the top of the siding with the soffit of the cornice.

frontage. N. The linear measurement of a piece of land along a lake, river, street, or highway.

front-end ratio. N. Lender calculation comparing a borrower's monthly housing expense (principal, interest, taxes, and insurance) to gross monthly income.

front footage. N. The number of feet of street frontage of a parcel of land.

front money. N. Amount of money necessary to start a project. Money invested in the initial stage of a business transaction to demonstrate good faith as well as to help offset some expenses.

frost line. N. The depth of frost penetration in soil and/or the depth at which the earth will freeze and swell. This depth varies in different parts of the country.

FSBO. ABBRV. For sale by owner.

full covenant and warranty deed. N. Type of property deed containing five warranties: covenant of seisin—assurance that the purchaser has possession of the property in quantity and quality as promised; covenant of quiet enjoyment—assurance of hostile claims to the property title; no liens and encumbrances; covenant of further assurance—assurance by the seller to the purchaser that all necessary actions to perfect the title will be undertaken, should any claims arise against the title; and, warranty of title.

full disclosure. N. Requirement to reveal any and all information pertinent to a transaction.

full recourse loan. N. A loan upon which an endorser or guarantor is responsible if the borrower defaults.

full-service broker. N. Real estate broker who performs all services including listing and selling.

full-service rent. N. Rent that includes operating expenses and real estate taxes for the first year.

full warranty. N. Warranty that entitles a homeowner to full remedies for defective work by a contractor.

fully amortized adjustable rate mortgage. N. Mortgage that pays down (amortizes) the balance of a loan.

fully amortized mortgage. N. Mortgage that has been paid in full and has no balance due.

fully depreciated. N. Of an asset, having already allocated the maximum allowable amount for the purposes of depreciation.

fully indexed rate. N. This rate is the interest rate that is used to calculate monthly payments in the absence of constraints imposed by the initial rate or caps. Fully indexed rate in conjunction with adjustable rate mortgages, the interest rate indicated by the sum of the current value of the index and margin applied to the loan.

functional deficiency. N. Negative characteristics about a property, which do not meet the needs of the usual occupant.

functional depreciation. N. Loss of value from all causes within the property, except for those due to physical deterioration.

functional modern/contemporary house. N. Post World War II style house with an exterior style that is an integral part of the overall design, incorporating modern technology, material, and architecture including energy conservation methods to achieve a highly functional structure.

funded debt ratio. N. The ratio of working capital from funds that have been borrowed to the working capital available from equity.

future advance clause. N. Clause in an open-ended mortgage that permits the mortgagor to borrow additional sums of money in the future by pledging the same real estate as collateral. Construction loans have a future advance clause providing additional loan guarantees as the building project progresses.

future interest. N. A property or estate right that may not be enjoyed until some time in the future when a certain event occurs, such as a life estate that will pass to another individual having a future interest as a fee simple estate.

future proposed space. N. In commercial real estate, a proposed development not yet under construction that is still waiting for a lead tenant, financing, zoning, approvals, or any other occurrence necessary to begin construction.

GAAP. ABBRV. Generally accepted accounting principles.

gable. N. Triangular wall enclosed by the sloping ends of a ridged roof and the top wall plate. The entire section, including the wall, roof, and space enclosed, is encompassed in the term.

gable roof. N. A ridged roof that forms a triangle at each end.

gag rule. N. A contract provision signed by new buyers that prohibits owners from making public any complaints about builders or other parties in a real estate transaction.

gain. N. The excess of the amount realized from a sale or exchange over the adjusted basis of the property sold or exchanged.

Gambrel roof. N. A roof with two slopes on each side, similar to a barn roof.

gap. N. Defect in the chain of title, such as a missing document, that raises doubt as to true ownership.

gap loan. See *bridge loan*.

garden apartment. N. (1) A below-ground-level apartment with a view of the lawn. (2) A particular type of housing project where all residents have access to a lawn area.

gated community. N. Exclusive, fenced-in development featuring a security guard at the entrance.

GDP. ABBRV. Gross domestic product.

GEM. ABBRV. Growing equity mortgage.

general contractor. N. Contractor who assumes responsibility for completing a construction project, under contract to the owner, and hires, supervises, and pays all subcontractors. A borrower using mortgage funds to renovate a home may be required by a

general depreciation system. N. The most commonly used modified accelerated cost recovery system; personal property is depreciated using the declining-balance method (double or 150%, depending on the recovery class), switching to straight line when that method results in the larger deduction. Residential rental property is depreciated using the straight-line method over 27.5 years, and nonresidential real property is depreciated using the straight-line method over thirty-nine years (31.5 years for property placed in service before May 13, 1993).

general lien. N. A lien that includes all of the property owned by the debtor rather than a specific property, and may be obtained either through a judgment lien, where the court issues a judgment; a lien by creditors on an estate; or, through federal and state tax liens.

generally accepted accounting principles. N. A widely accepted set of rules, conventions, standards, and procedures for reporting financial information, as established by the Financial Accounting Standards Board. ABBRV. *GAAP*.

general partner. N. The managing partner of a limited partnership who is in charge of its operations and has unlimited liability. All partners in an ordinary partnership are general partners. A limited partnership must have one general partner.

general plan. N. A government's long-range land-use plan.

general straight-line depreciation system. N. A modified accelerated cost recovery system (MACRS) of depreciation using the straight-line method over the normal MACRS recovery period for the asset.

general warranty deed. N. A deed in which the grantor agrees to protect the grantee against any other claim to title of the property. The covenants assure good title, freedom from encumbrances and quiet enjoyment.

geodesic dome. N. A structure constructed of lightweight bars forming a grid of polygons.

Georgian. ADJ. A large, English-style home.

gift tax. N. Federal tax placed on a gift, monetary or property. The tax is based on the appraised value (if other than monetary) at the time of transfer. Under current tax law, each person may gift up to $11,000 tax free annually to another person.

G.I. loan. N. A loan made through the Department of Veterans Affairs.

G.I. loan guarantees. N. See *VA guarantee*.

gingerbread decoration. N. Lacy, intricate wooden trim, like that on a gingerbread house.

Ginnie Mae. ABBRV. Government National Mortgage Association.

girder. See *beam*.

GLA. ABBRV. Gross leasable area.

GNMA. ABBRV. Government National Mortgage Association.

good and marketable title. See *clear title*.

good faith. N. Concept that each party in a real estate transaction is presumed honest and fair with no deceit and that their intentions are honorable and realistic. If deception occurs, without prior knowledge, the transaction, carried out in good faith, remains valid.

good faith estimate. N. A report from a lender that outlines the costs a borrower will incur to get a mortgage. It is based on the lender's typical loan origination costs for the area where the home is located. The estimate usually changes between application and closing, so borrowers should review the settlement form before the closing meeting. The settlement form will list the actual amount of money the borrower will need to bring to closing. Closing costs will need to be paid with a certified or cashier's check because personal checks usually are not accepted.

good repair clause. N. Contract clause indicating that the property must be properly maintained to keep the contract valid, which creates liability for the seller or lessee if the subject property is found to be in need of repairs.

goodwill. N. The ability of a business to generate income in excess of a normal rate on assets due to superior managerial skills, market position, new product technology, etc. In the purchase of a business, goodwill represents the difference between the purchase price and the value of the net assets. Goodwill acquired after August 10, 1993, must be amortized over a fifteen-year period and is subject to recapture when the business is sold.

government mortgage. N. A mortgage that is insured by the Federal Housing Administration or guaranteed by the Department of Veterans Affairs or the Rural Housing Service.

Government National Mortgage Association. N. A government-owned corporation within the U.S. Department of Housing and Urban Development. Created by Congress on September 1, 1968, the Government National Mortgage Association assumed responsibility for the special assistance loan program formerly administered by Fannie Mae. Popularly known as Ginnie Mae, the corporation funds high-risk mortgages typically in areas approved for government construction projects that have no other funding sources. The government body also buys home loans issued by others, such as commercial banks, mortgage banks, and insurers and, after pooling them together, sells shares to investors. Unlike Fannie Mae and Freddie Mac, Ginnie Mae is backed by the United States and thus has a higher credit standing. ABBRV. *GNMA, Ginnie Mae.*

government rectangular survey. See *rectangular land survey.*

GPM. ABBRV. Graduated payment mortgage

grace period. N. Period of time during which a loan payment may be made after its due date without incurring a late penalty. The grace period is specified as part of the terms of the loan in the lending agreement.

grade. N. The elevation of land above level ground.

grade beam. N. A foundation wall that is poured at level with or just below the grade of the earth.

graded tax. N. Increasing tax rates as levels of taxable income rise.

grade level. N. The flat or sloping surface upon which a house is built.

graduated payment mortgage. N. Mortgage where the monthly payments are low for the first few years, gradually rise for a few years, and then remain fixed. ABBRV. *GPM.*

Graduate Realtor Institute. N. Designation issued by the National Association of Realtors® (NAR) to members meeting specific performance and education requirements for residential real estate. ABBRV. *GRI.*

granny flat. N. Expression for a separate apartment unit within a house or above the garage to either rent out or that might be shared by a relative.

grant. N. The technical term used in a deed of conveyance of property to indicate a transfer to another party.

grantee. N. One to whom an interest in a piece of property is conveyed.

grantor. N. Person conveying interest in a piece of property to another.

grantor/grantee index. N. A reference kept with public records that cross-indexes grantors and grantees with one another, along with the properties to which they relate.

Greek revival style. N. Style whose most prominent feature is a pillar-anchored pediment forming a portico in front of the house.

Greek revival style

greenbelt. N. Any area of park land, open space, or other public nature preserve within a community.

GRI. ABBRV. Graduate Realtor Institute.

grid. N. (1) A pattern of lines laid out at right angles to each other. (2) A series of intersecting lines dividing a map or chart into equal sections. (3) The intersecting bars, wires, or supports as in a grating or a dropped ceiling.

gross area. N. The total floor area of a structure, in square feet, measured from the outside.

gross domestic product. N. Measurement of the value of all goods and services produced by the economy within its boundaries. Gross domestic product (GDP) is normally stated in annual terms, though data is compiled and released quarterly. The government gives a preliminary figure every quarter and revises it twice. GDP is often a measure of the state of the economy. ABBRV. *GDP*.

gross income. N. (1) Total income of a household before expenses and taxes are subtracted. (2) Cash received from a room rental.

gross income multiplier. N. Method used to compute the price of an income-producing property by dividing the asking or market price of the property by the current gross rental income. If the current gross rental income is $30,000 and the asking price is $300,000, the gross income multiplier is ten. Synonymous with *gross rent multiplier, rent multiplier*.

gross leasable area. N. A building's total floor area, in square feet, designed for tenant leasing. ABBRV. *GLA*.

gross lease. N. Rental in which the lessor pays all operating costs such as taxes, utilities, insurance, and maintenance, in addition to the rent.

gross margin. N. Gross income divided by net sales, expressed as a percentage.

gross profit. N. Profit remaining after the deduction of direct costs but before the deduction of expenses.

gross rent multiplier. See *gross income multiplier*.

gross rents. N. Total income from rents before expenses or the depreciation or cost recovery deduction.

gross ups. N. The adjustments made to operating expenses in a commercial or residential building to reflect their actual occupancy levels.

ground fault circuit interrupter. N. Safety device that detects the leakage of electrical current in the ground.

ground lease. N. Lease of land only.

ground plan. N. View of a plot showing the structures located upon it. Ground rent portion of property income earned by the leasing value of the land.

ground rent. N. The amount of money paid for the use of a piece of property when it is a leasehold estate.

group home. N. A single-family residential structure designed or adapted for occupancy by unrelated developmentally disabled persons. The structure provides long-term housing and support services that are residential in nature.

growing equity mortgage. N. A fixed-rate mortgage that provides scheduled payment increases over an established period of time, with the increased amount of the monthly payment applied directly toward reducing the remaining balance of the mortgage. ABBRV. *GEM*.

guarantee. N. A financially binding guarantee assuring that the guarantor will fulfill an obligation or contractual agreement.

guaranteed payment loan. N. Assurance that a loan's financial obligation will be secured by a third party.

guaranteed sales program. N. Real estate brokerage program that purchases the seller's equity if a property does not sell during a certain period of time.

guarantee mortgage. N. Loan that is guaranteed by a third party; for example, a government institution.

guaranty. See *guarantee*.

guardian. N. One who is appointed to administer the personal affairs and property of an individual who is incompetent.

guardian's deed. N. Deed used to convey property of a minor or legally incompetent person.

gutter. N. A shallow metal channel attached to a structure that carries rainwater from the roof.

gypsum board. N. Drywall, wallboard, or gypsum.

habendum clause. N. Clause that defines or limits the quantity of the estate granted in the deed. Declares whether the type of ownership conveyed is fee simple, a life estate, or something different. Synonymous with *to have and to hold clause*.

half-bath. N. A bathroom with a toilet and a washbasin but no shower or tub.

handyman's special. N. In real estate advertising, this generally refers to a property that requires significant renovation, though sells at an attractive price.

hard money. N. (1) Currency that has wide acceptance, such as the U.S. dollar. (2) Gold or silver coins, as compared to paper currency. (3) Actual cash exchanged in a loan; sometimes used to describe extremely high-interest-rate mortgage loans made to desperate borrowers.

hazard. N. Condition that affects the probability of losses or perils occurring, such as flood damage to a house.

hazard insurance. N. See *homeowners' insurance*.

hazardous waste site. N. Property or land that contains hazardous waste; the Environmental Protection Agency has identified contaminated hazardous waste sites across the country. See also *contingency*.

H-clip. N. Small metal clip formed like an "H" that fits at the joints of two plywood (or wafer board) sheets to stiffen the joint. Normally used on the roof sheeting.

header. N. Crossbeams above windows and doors.

hearth. N. The fireproof area directly in front of a fireplace.

heat pump. N. An electric cooling and heating system.

heavy timber construction. N. Use of heavy timbers, connected with bolts and metal plates at their intersections, for main structural pieces in construction. The heavy timbers carry the structural load so that studs are added to form partitions and not for weight bearing.

HECM. ABBRV. Home equity conversion mortgage.

hectare. N. A measurement equaling 2.471 acres or about 107,637 square feet or 10,000 square meters.

heir. N. Individual legally entitled to inherit money and property on the death of another person.

heirs and assigns. N. Language commonly used in a fee simple title conveyance. The significance is whether the title is clear and can be passed on to the purchaser's estate, including all heirs and those who may have any interest in the estate, i.e., the assigns.

HELOC. ABBRV. Home equity line of credit.

hereditaments. N. Property, real estate or personal, tangible or intangible, that may be inherited.

heterogeneous. N. (1) Mixed assortment of housing styles in a residential development. (2) Mixed zoning uses in an urban development plan.

hiatus. N. Gap between two parcels of land, which is not included in the legal description of either parcel.

hidden asset. N. Asset not immediately apparent from a balance sheet.

hidden clauses. N. Ambiguous contractual language that may result in an unsuspecting buyer of real estate incurring obligations or risks not clearly evident.

high density

high density. N. The concentration of housing units on a specific property or in a specific area.

highest and best use. N. Appraisal term meaning the legally and physically best possible use that will produce the greatest current value.

high loan-to-value loan. N. A loan covering more than 100% of the market value of the home. Such loans are used as a refinancing tool, essentially making them home equity loans.

high-rise. N. A building usually taller than six stories and serviced by elevators. The designation is determined by local codes.

highway easement. N. The construction of a highway right of way over a privately held parcel of land. Property owners are compensated for the value of the property usurped by a highway easement.

hip roof. N. A pitched roof with sloping sides.

historical cost. N. An accounting principle requiring all financial statement items to be based on original cost.

historic district. N. Area designated by government to have historical importance. Various incentives, including tax breaks to rehabilitate and preserve the area, are provided.

historic preservation. N. (1) A movement begun in the 1960s in the United States to protect landmarks and to unify neighborhoods. (2) The physical rehabilitation of a historic building.

historic structure. N. (1) A building that is listed in the National Register of Historic Places and certified as historic by the U.S. Secretary of the Interior. A building that is officially recognized for its historic significance has special status under the 1997 Tax

hold back. N. Portion of a construction loan withheld by a lender from a contractor until all construction work is satisfactorily completed or sufficient space is leased in a floor loan.

holder in due course. N. (1) Legal ruling providing protection to homebuyers of defective homes bought from a seller who then sold the contract to a third party. (2) One who acquires a bearer instrument in good faith and is eligible to keep it even though it may have been stolen.

hold harmless clause. N. Contractual clause where one party assumes a liability risk for another and, thus, effectively indemnifies the named party from any liability.

holding company. N. Company formed for the purpose of owning or controlling other companies.

holding funds. N. Funds retained in an account until a certain event occurs.

holding period. N. The period of time property has been owned for income tax purposes. The holding period determines if gain or loss from the sale or exchange of a capital asset is long- or short-term.

Home Affordability Index. N. Measure of the typical U.S. family's ability to buy a home, published by the National Association of Realtors®. When the index measures one hundred, a family earning the median income has exactly the amount needed to purchase a median-priced, previously owned home using conventional financing and a 20% down payment. Some experts say that every 1-point increase in the home mortgage interest rates results in three hundred thousand fewer home sales.

home equity conversion mortgage. N. A loan made to older owners—62 and over—to convert their equity into money. Borrowers are qualified on the basis of the value of their homes. In addition, the loan does not have to be repaid until the borrower no longer occupies the property. The equity can be paid to the homeowner in a lump sum or in a stream of payments, drawn from a line of credit, or a combination of monthly payments and line of credit. The advantages are that there are no restrictions on how the funds should be spent, the funds do not affect Social Security or Medicare benefits, and any remaining equity at the time of the borrowers' death may be passed on to heirs. The requirements are that borrowers must be 62 or older and must have their home mostly paid off. Eligible properties include a single-family home, a two- to four-unit dwelling, a condominium, or a manufactured home. All housing types must meet Federal Housing Administration guidelines. This home must be the borrower's principal residence. In general, a borrower can get between one-third and one-half of total equity as a line of credit or as a lump sum payment. ABBRV. *HECM.* Synonymous with *reverse mortgage.*

home equity line of credit. N. Open-ended line of credit based on a homeowner's equity. Most loan amounts are limited to 75% or 80% of the appraised value. Withdrawals can be made at any time with the guidelines. ABBRV. *HELOC.*

home equity loan. N. Loan allowing owners to borrow against their equity in the home; usually a second mortgage.

home improvement loan. N. Loan used to pay for major remodeling, reconstruction, or additions to the home; usually a second mortgage.

home inspection. N. Examination of a home's condition, internal systems, or construction prior to purchase—this should be done by a professional contractor or an appraiser knowledgeable about:

roofs and siding; windows and doors; foundation; insulation; ventilation; heating and cooling systems; plumbing and electrical systems; walls, floors, and ceilings; and, common areas in a condominium or cooperative. Prospective buyers should view the home inspection report as a way to identify problems before purchasing to help negotiate adjustments in the purchase price if problems exist and to help get the buyer to make any needed improvements before the deal is finalized. See also *contingency*.

home inspector. N. Professional who does home inspections and evaluates the structural soundness and operating systems of a home.

home inventory. N. Listing of items and their costs of an individual's possessions at his or her residence.

home loan. N. See *mortgage*.

home office expenses. N. Expenses of operating a portion of a residence used for business- or employment-related purposes.

homeowners' association. N. Group that governs a planned community or condominium and collects monthly fees from all owners to pay for common area maintenance, handle legal and safety issues, and enforce the conditions and restrictions set by the developer. ABBRV. *HOA*.

homeowners' fee. N. Fee charged to a homeowner to belong to a homeowners' association, which includes the cost of maintenance and other services.

home ownership. N. The state of living in a structure that one owns.

homeowners' insurance. N. Type of insurance policy covering the risks of homeowners, including damage, theft, fire, personal liability, etc. Homeowner's insurance should be equal to at least the replacement cost of the property. Replacement cost coverage ensures that a home will be fully rebuilt in case of a total loss. Most homebuyers purchase a homeowners' insurance policy that includes personal liability insurance, in case someone is injured on their property; personal property coverage for loss and damage to personal property due to theft or other events; and, dwelling coverage to protect the house against fire, theft, weather damage, and other hazards. If the home is located near water, a lender might require flood coverage. This insurance is generally expensive. Lenders often want the first year's insurance premium to be paid at or before closing—sometimes a lender may require this payment to be made monthly in a required escrow account. Synonymous with *hazard insurance*.

Homeowners Protection Act of 1997. N. Requires private mortgage companies to tell borrowers of their right to cancel mortgage insurance when the loan amount is no more than 80% of the value of the home. Covers loans originated after July 31, 1999.

homeowners' warranty. N. A type of insurance that covers repairs to specified parts of a house for a specific period of time. It is provided by the builder or property seller as a condition of the sale. ABBRV. *HOW*.

homeowners' warranty program. N. Private insurance program that protects purchasers of newly constructed homes against structural and mechanical defects, and provides reimbursement for the cost of remedying the situation if the builder does not do so.

home price. N. Price agreed upon by seller and purchaser and for which title is exchanged.

home rule. N. Power of the local government to implement its own land-use regulations.

homestead. N. Legal status provided by certain states on a homeowner's principal, which in some states provides protection against creditor claims or forced land sale as long as the homeowner continues to maintain his or her residence there.

homestead law. N. Law that exempts a homestead from forced sale to meet general debts.

home warranty. N. Warranty issued by contractors, sellers, and real estate agencies that protects homebuyers from specified defects in a house as per the contract.

homogenous. N. Something constructed with parts of the same material. ADJ. Term for an area where property types and uses are similar and compatible.

hopper window. N. A window that contains a single sash that tilts inward.

horizontal property law. N. Body of law relating directly to condominiums and cooperative developments. Most property law provides vertical ownership of property in the sense that property owners own mineral rights as well as air rights to property. Horizontal property laws state that property owners own only the confines of the apartment unit within a condominium or cooperative building complex. Thus, horizontal property laws do not allow property owners to own the land on which their apartment unit is located.

hose bibb. N. A threaded faucet connection for devices such as a washing machine. Incorrect connections can cause flooding.

hours of operation. N. In a commercial lease, the hours of operation for a property are specified as a way to measure use and risk.

household employee. N. An individual who performs nonbusiness services for the taxpayer in or around the taxpayer's home. Such services include child and dependent care, house cleaning, cooking, and yard work.

household expenses. N. A portion of total support; the value of lodging plus food consumed in the home, utilities paid, and repairs made. The total is divided equally among all family members. Each member's share of household expenses is part of his or her total support.

house poor. N. Purchasing a more expensive house than a buyer can afford based on his or her income.

house wrap. N. A polyethylene barrier wrapped around a house to save energy.

housing code. N. Federal, state, or local government ordinance that sets minimum standards of safety and sanitation for existing residential buildings, as opposed to building codes, which govern new construction.

housing discrimination. N. Illegal practice of denying the right to buy or rent a home to an individual based on race, religion, color, national origin, sex, disability, or family status.

housing expense ratio. N. The percentage of gross monthly income that goes toward paying housing expenses.

housing finance agency. N. Financial agency typically associated with state or local governments. These agencies are generally geared toward assisting first-time and low- to moderate-income borrowers. They use tax-exempt bonds to fund mortgage lending, and as a result are often able to provide interest rates that are below current market rates.

housing starts. N. Estimate of the number of dwelling units on which construction has begun during a stated period.

HOW. ABBRV. Homeowners' warranty.

HUD. ABBRV. U.S. Department of Housing and Urban Development.

HUD-1 Settlement Statement. N. The HUD-1 Settlement Statement itemizes the amounts to be paid by the buyer and the seller at closing. Items on the statement include real estate commissions, loan fees, points, and escrow amounts. The form is filled out by the closing agent and must be signed by the buyer and the seller. The buyer should be allowed to review the HUD-1 Settlement Statement on the business day before the closing meeting to know the closing costs in advance. Synonymous with *closing statement, settlement sheet, settlement statement*.

HUD median income. N. Median family income for a particular county or metropolitan statistical area as estimated by the Department of Housing and Urban Development.

humidifier. N. An appliance attached to the furnace, or a portable unit used to increase the humidity within a room.

HVAC. ABBRV. Heating, ventilation, and air conditioning; used to refer to climate control systems.

hybrid method of accounting. N. A combination of accounting methods, usually of the cash and accrual methods.

I-beam. N. A steel beam with a cross section resembling the letter "I." It is used for long spans as a basement beam or over a wide wall opening, such as a double garage door, when wall and roof loads bear down on the opening.

illiquid. ADJ. Cannot be quickly sold or converted to cash without incurring a significant loss. Real estate is generally an illiquid investment.

immunization. N. Protection against interest rate risk by holding assets and liabilities of equal durations.

impact fees. N. Fees that must be paid by developers of new homes and subdivisions to pay for town facilities such as schools and parks.

impaired credit. N. Decline in the credit status of a prospective borrower.

implied condition. N. A provision not explicitly stated in an agreement but considered an important item.

implied contract. N. An agreement created by actions of the parties involved but not written or spoken.

implied easement. N. Property that is used consistently for many years without challenge by the actual owner.

implied warranty. N. Under law, there is an express warranty that real estate sold is appropriate for sale and is in proper condition, even if not stated.

implied warranty of habitability. N. Legal doctrine that all new homes are assumed to meet all building codes and are fit for habitation.

impound account. N. See *escrow account*.

improved land. N. Land that has been developed for use and has had installation of such utilities as water, sewer, roads, and building structures. These improvements make the raw land increase its usability, thereby increasing the market value.

improvement. N. Change to a house that adds value, prolongs its use, or adapts it to different use.

improvement ratio. N. The relative value of improvements to the value of the original, unimproved property.

imputed interest. N. In the case of certain long-term sales of property, the Internal Revenue Service has the authority to convert some of the gain from the sale into interest income if the contract does not provide for a minimum rate of interest to be paid by the purchaser. Such converted interest is called imputed interest. Synonymous with *unstated interest*.

imputed value. N. The value of an asset that is not recorded in any accounts but is implicit in the product. For example, when making historical comparisons, an imputed value can be estimated for any period for which data is not available.

inactive. ADJ. Not continuously in use, e.g., a thinly traded security.

inactive asset. N. An asset that is sometimes not continuously utilized, such as a backup power generator or a secondary system used only when the primary system malfunctions.

incandescent lamp. N. A lamp employing an electrically charged metal filament that glows as white heat; a typical light bulb.

income. N. Money or other benefits coming from the use of property, skill, or business; the excess of revenue over expenses and losses for an accounting period. For purposes of the passive loss rules, income must be divided into three categories: active

income approach

income, passive income, and portfolio income. Synonymous with *earnings*.

income approach. N. Method of appraisal for real estate based on the property's anticipated future income; market value equals expected annual income divided by the capitalization rate.

income-producing property. N. Investment property; real estate held for investment potential or in order to earn income by leasing or letting it, rather than for its own use.

income property. N. Property that is used to generate income, i.e., rented to others as either commercial or residential.

income statement. N. Profit and loss statement; financial statement depicting a business entity's operating performance and reports the components of net income, including sales of real estate, rental income, operating rental expenses, income from rental operations, and income before tax. The income statement shows the cash flow for an entire accounting period, usually a quarter. The income statement is included in the annual report of the real estate corporation. Synonymous with *earnings report, profit and loss statement*.

income stream. N. A regular flow of money generated by a business or investment.

income yield. N. See *capitalization rate*.

incompetent. N. One who is not legally capable of completing a contract. This includes the mentally ill and minors. ADJ. Incapable of performing duties because of a lack of knowledge and training.

incorporate. N. To form a corporation under state regulations provided by the secretary of state.

incorporeal property. N. Legal interests and rights in real estate that do not include the right of possession, such as air and mineral rights, riparian rights, easements, and access rights.

incurable defect. N. A defect in a property that cannot be fixed or is too expensive to repair.

incurable depreciation. N. When the cost of repairing a component of a structure exceeds the value of the structure, making it economically impractical to repair.

indemnify. V. To protect another person against loss or damage or to compensate a party for loss or damage.

indenture. N. Written agreement between two or more persons having different interests.

independent appraisal. N. Value estimate provided by someone who has no participation in ownership of the property in question.

independent auditor. N. A Certified Public Accountant who provides a company with an accountant's opinion but who is not otherwise affiliated with the company.

independent contractor. N. A taxpayer who contracts to do work according to his or her own methods, and who is not subject to control except as to the results of such work. An employee, by contrast, is subject to the control of the employer as to the methods to be used to obtain the desired results.

index. N. A number used to compute the interest rate for an adjustable rate mortgage (ARM). The index is generally a published number or percentage, such as the average interest rate or yield on Treasury bills. A margin is added to the index to determine the interest rate that will be charged on the ARM. This interest rate is subject to any caps that are associated with the mortgage.

indexed loan. N. A long-term loan in which the term, payment, interest rate, or principal amount may be adjusted periodically according to a specific index, which is usually stated in the loan agreement.

index lease. N. A rental contract in which the tenant's rental is tied to a change in the price level, such as the gross national price deflator.

Index of Leading Economic Indicators. N. This index indicates the direction of the economy in the next six to nine months and helps to forecast business trends. This series of eleven indicators is calculated and published monthly by the U.S. Department of Commerce.

index of residential construction cost. N. Index of the costs to construct residential properties.

indirect costs. N. Costs not directly associated with the structure itself but incurred during the construction period. Synonymous with *soft costs*.

indirect overhead. N. Costs that are not related specifically to one particular job but are a general cost of doing business.

individual net worth. N. Total assets less total liabilities less estimated taxes equals an individual's personal equity, which is normally the basis upon which a loan is given.

industrial park. N. An area zoned and planned for the purpose of industrial development. Usually located outside the main residential area of a city and normally provided with adequate transportation access, including roads and railroad.

industrial property. N. Property that is zoned and used for industrial purposes, such as factories, manufacturing, research and development, warehouse space, and industrial parks.

industrial tract. N. Land zoned for industrial use, such as manufacturing, factory office and warehouse space, and research and development.

industrial zoning. N. Category of property zoning that designates property to be used for industrial purposes.

in escrow. N. Phrase used for the period in which the escrow agent communicates to both the buyer and the seller as to what documents or monies have to be deposited with the escrow agent to satisfy the terms of the purchase and sale. The items collected include monies to cover mortgage insurance premiums, taxes, hazard insurance, title insurance, termite inspection certificate, the seller's original deed, and confirmation (i.e., commitment) that the seller has obtained an adequate loan to pay the purchase price.

in-file credit report. N. Computer-generated report from credit repositories, which is regarded as an objective history.

infill development or housing. N. New construction in an already established area.

infiltration. N. The passage of air from indoors to outdoors and vice versa; term is usually associated with drafts from cracks, seams, or holes in buildings.

inflation. N. An increase in the amount of money or credit available in relation to the amount of goods or services available, which causes an increase in the general price level of goods and services. Over time, inflation reduces the purchasing power of a dollar, making it worth less.

inflation accounting. N. Showing the effects of inflation on financial statements, a Financial Accounting Standards Board requirement for large companies.

inflation equity. N. Increase in the value of property brought on by inflation.

information reporting. N. Income reporting to the Internal Revenue Service using Form 1099 stating income earned.

information returns. N. These are returns, such as Form W-2 and the various 1099 forms, that report income and property transactions to the Internal Revenue Service. The payer, broker, or other designated person is required to file these returns and is subject to penalties for noncompliance.

infrastructure. N. The basic public works in a city including roads, parks, bridges, schools, utilities, and communication systems in a community.

ingress. N. (1) Access from a land parcel to a public road or other means of entrance. (2) Right to enter through land owned by another.

inheritance. N. As distinguished from a bequest or devise, an inheritance is property acquired through laws of descent and distribution from a person who dies without leaving a will. Property so acquired usually takes as its basis, for gain or loss on later disposition or for depreciation, the fair market value at the date of the decedent's death. An inheritance of property is not a taxable event, but the income from an inheritance is taxable.

inheritance tax. N. State tax based on the value of property received through inheritance.

initial interest rate. N. The original interest rate of the mortgage at the time of closing. This rate changes for an adjustable rate mortgage (ARM). See also *teaser rate*.

initial payment. N. The down payment on the price of a piece of real estate.

initial rate. N. The rate charged during the first interval of an adjustable rate mortgage.

initial rate cap. N. Specific limit of some adjustable rate mortgages defining the maximum amount the interest rate may increase at the expiration of the original interest rate.

initial rate duration. N. The time period, lasting either months or years, before the initial interest rate expires and an increase takes place.

injunction. N. A court order issued to a defendant in an action either prohibiting or commanding the performance of a defined act. Violation of an injunction could lead to a contempt of court citation.

in re. PREP. (*Latin*) Concerning; in the matter of.

in rem. ADJ. (*Latin*) Against the thing; legal term describing proceedings against property rather than proceedings against people. ADV. *in rem*.

insolvent. N. A financial condition in which a taxpayer's total liabilities exceed the total fair market value of all his or her assets. A taxpayer is insolvent to the extent his or her liabilities exceed his or her assets.

inspection. N. Examination or review, which compares objects by use of an acceptable standard to rate quality and to guarantee consistency.

inspection fee. N. Fee paid to a professional to determine the physical condition of a home to supplement the information in the appraisal report and is often required by the lender.

inspection report. N. The written report submitted after an inspection detailing the condition of the home's foundation, framing, plumbing, electrical system, heating, air conditioning, fireplaces, kitchen, bathrooms, roof, exterior, and interior.

inspector. N. A person who inspects; an official examiner.

installment. N. Any of several parts appearing at intervals; periodic payments of debt in equal parts.

installment contract. N. Purchase agreement where the buyer does not receive title to the property until all installments are paid.

installment loan. N. Borrowed money that is repaid in equal payments, known as installments. A furniture loan is often paid for as an installment loan.

installment method. N. A method of accounting that enables a taxpayer to spread the recognition of gain on the sale of property over the payment period. Under this procedure, the seller computes the gross profit percent from the sale (that is, the gain divided by the contract price) and applies it to each payment received to arrive at the amount of the gain to be recognized.

institutional lender. N. Financial intermediary that invests in loans and other securities on behalf of its depositors or customers.

instrument. N. Written legal document.

insulating glass. See *double glass*.

insulation. N. Materials including cellulose, glass fiber, rock wool, polystyrene, urethane foam, and vermiculite that slow heat loss.

insurable interest. N. An interest in a person or property that would cause one a loss if that person or property were injured or ruined.

insurable title. N. Title to a property that can be insured against defects and disputes.

insurance. N. Policies that guarantee compensation for losses from a specific cause. Various forms of insurance cover against fire, flood, earthquake, liability, etc.

insurance binder. N. Temporary insurance arrangement used until a permanent policy can be issued.

insurance company. N. An organization that underwrites insurance policies.

insurance dividend. N. An amount paid to policyholders; this is not a dividend on capital stock, but a rebate of a portion of the premiums paid for the insurance. Such dividends reduce the cost of the insurance and are not taxable unless in excess of the total premiums paid. Interest paid when the dividends are left with the insurance company is reported to the taxpayer as interest and is taxable.

insured mortgage. N. A mortgage that is protected by the Federal Housing Administration or by private mortgage insurance. If the borrower defaults on the loan, the insurer must pay the lender the lesser of the loss incurred or the insured amount.

intangible asset. N. Nonphysical valuables, such as contracts or mortgages, or employee loyalty or customer goodwill, that are distinguished from physical property such as buildings and land.

intangible personal property. N. Property other than real property with no intrinsic value; its value lies in the rights conveyed. Examples include cash, insurance, stock, goodwill, and patents.

interest. N. Charge paid for borrowing money, calculated as a percentage of the remaining balance of the amount borrowed.

interest accrual rate. N. The percentage rate at which interest accrues on the mortgage. In most cases, it is also the rate used to calculate the monthly payments, although it is not used for an adjustable rate mortgage with payment change limitations.

interest cover. N. The annual rate of interest on the loan, expressed as a percentage of one hundred.

interest-only loan. N. Loan for which only the interest is paid each month. Therefore, the outstanding amount of principal is not reduced.

interest rate buydown plan. N. An arrangement wherein the property seller (or any other party) deposits money to an account so that it can be released each month to reduce the mortgagor's monthly payments during the early years of a mortgage. During the specified period, the mortgagor's effective interest rate is bought down below the actual interest rate.

interest rate cap. N. Consumer safeguards that limit the amount the interest rate on an adjustable rate mortgage loan can change in an adjustment interval and/or over the life of the loan. For example, if a per-period cap is 1% and the current rate is 7%, then the newly adjusted rate must fall between 6% and 8%, regardless of actual changes in the index.

interest rate ceiling. N. Highest interest rate allowed to be charged on an adjustable rate mortgage.

interest rate floor. N. The minimum interest rate for an adjustable rate mortgage, as specified in the mortgage note.

interest rate for HECMs. N. The interest rate on a home equity conversion mortgage (HECM) adjusts monthly or yearly. It is tied to the weekly average yield of U.S. Treasury securities adjusted to a constant maturity of one year. The interest charged on the HECM loan will be payable to the lender when the loan terminates.

interest received. N. An amount received for the use of money that is to be repaid in full at a specified time or on demand.

interim financing. N. See *bridge loan*.

interim loan. N. Loan that is to be replaced by a permanent loan.

interim statement. N. Financial report covering less than one year, such as a quarterly report.

interim use. N. Temporary use of property in a nonconforming use, which can be overturned by a formal zoning ruling.

interior finish. N. Material used to cover the interior framed areas of walls and ceilings.

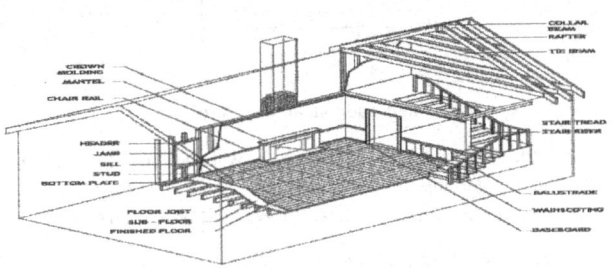

Interior finish

intermediation. N. Court order on an issue directly related to the immediate action.

internal rate of return. N. Real annual return on a real estate investment. It equates the initial investment with the present value of future net cash inflows from the investment. ABBRV. *IRR*.

Internal Revenue Code. N. U.S. tax law consisting of rules and regulations to be followed by taxpayers, which is continually revised and amended. ABBRV. *IRC*.

Internal Revenue Service. N. The branch of the federal government responsible for collecting taxes, including corporate and personal. It also administers tax rules and regulation and investigates tax irregularities. ABBRV. *IRS*.

international architecture. N. A design of architecture that originated in Europe in the 1920s; its design was very functional and emphasized buildings constructed of steel, reinforced concrete, and glass. Smooth white surfaces with very large wall windows and no decoration advertised that functionality was important and "less is more"; it was possibly a reaction to the highly decorated houses of the Victorian period.

International architecture

intestate. N. Person who dies without a will and having unknown intentions regarding the disbursement of his or her estate. In this case, a court-appointed administrator typically acts as an executor.

in toto. ADV. (*Latin*) In the hole or as a hole.

inventory. N. Property held for sale or to be used in the manufacture of goods held for sale.

investment. N. (1) Expenditure to buy property or other capital assets that generate income. (2) Securities of real estate companies or capital assets.

investment analysis. N. Analysis of the risks and rewards to an individual in making a particular property investment. It considers the cost of the original investment, the investment return over a period of time, the suitability of the investment, and the probability of success.

investment banks. N. Large institutions that deal mainly with underwriting, raising capital, securities, corporate mergers and acquisitions, and investing.

investment flows. N. Cash flows associated with the buying and selling of fixed assets and business interests.

investment interest. N. Interest paid on loans acquired to purchase or hold investment property. Investment interest is deductible as an itemized deduction to the extent of net investment income.

investment life cycle. N. Time interval between buying a real estate investment and selling it.

investment property. N. Real estate, such as rental properties, that generates income.

investment return. N. What an investor yields on the investment in dollars as a percentage.

invoice. N. A bill issued by one who has provided products and/or services to a customer. In asset-based lending, invoice means "account receivable."

involuntary alienation. N. Loss of property due to attachment, condemnation, foreclosure, sale for taxes, or other involuntary transfer of title.

involuntary conversion. N. The receipt of money or other property as reimbursement for the loss or destruction of property through theft, casualty, or condemnation. Any gain realized on an involuntary conversion can, at the taxpayer's election, be considered nonrecognizable for federal income tax purposes if the owner reinvests the proceeds within a prescribed period of time in similar property.

involuntary lien. N. Lien on property such as for the nonpayment of real estate taxes or a mechanic's lien.

I persona. N. (*Latin*) I am the person; meaning that it is actually the person him- or herself.

ipso facto. ADV. (*Latin*) By the act or fact itself.

IRC. ABBRV. Internal revenue code.

IRR. ABBRV. Internal rate of return.

irrevocable. ADJ. Cannot be taken, returned, or revoked.

irrigation system. N. Lawn sprinkler system.

IRS. ABBRV. Internal Revenue Service.

Italian architecture. N. Style of architecture introduced in America prior to the Civil War, modeled after Renaissance country homes in northern Italy. Homes in this style were usually relatively large brick houses, which were characterized by having an off-center square tower and a flat roof with heavy overhanging eaves supported by braces.

Italian architecture

itemized deductions. N. Certain personal expenditures allowed by the tax code as deductions from adjusted gross income. Examples are certain medical expenses, qualified interest on home mortgages, and charitable contributions. Itemized deductions are reported on Schedule A, Form 1040. A taxpayer who itemizes deductions may not claim the standard deduction.

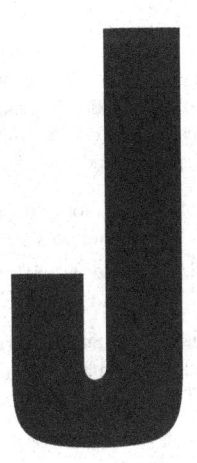

jalousie window. N. A window with vertical rows of horizontal glass slats that opens and closes with a crank mechanism that connects all the slats.

jamb. N. The lining of a doorway.

J Channel. N. Metal edging used on drywall to give the edge a better finished appearance.

jeopardy. N. Danger, risk factor.

J Factor. N. Factor used by appraisers and investors to determine the changes needed in operating income to obtain a desired rate of return. This factor is determined after consideration of the location, similar properties, and cost of maintenance.

joint. N. A union of two or more people who are either for or against something. ADJ. (1) Indicates a common property ownership interest in real estate. (2) Indicates a shared liability in terms of a contractual relationship.

joint and several liability. N. Situation wherein each borrower, on the same note, is held fully liable for the entire amount of the debt, not just a portion. The creditor may demand full repayment from any individual.

joint liability. N. Situation where two or more people share the responsibility of fulfilling the terms of a debt.

joint note. N. Note on which there is more than one maker; if one or more of the makers default on the note, all of the makers are sued jointly, rather than just one or all, to make restitution.

joint ownership. N. Ownership of real estate with two or more individuals having equal ownership in which, upon the death of one owner, the property is transferred to the survivor.

joint tenancy. N. Equal ownership by two or more people, each of whom has an undivided interest, with the right of survivorship.

joint venture. N. An agreement between two or more parties to invest in a specific single business or property. Although not a continuing relationship, it is treated as a partnership for income tax purposes.

joist. N. A floor or ceiling support member supported by foundation walls, piers, or beams. Subflooring is connected to floor joists.

journal. N. An accounting book of original entry where transactions are initially recorded.

journeyman. N. Worker who has already served his or her apprenticeship to work in a trade such as plumbing or carpentry, but who is not yet a supervisor.

judgment. N. A decision made by a court of law. In judgments that require the repayment of a debt, the court may place a lien against the debtor's real property as collateral for the judgment's creditor. Synonymous with *adjudication order*, *decree*.

judgment creditor. N. The party to whom the court awarded a financial judgment against a debtor.

judgment debtor. N. The party against whom the court has placed a financial judgment.

judgment lien. N. Court order in which the judgment creditor is granted a lien against the property of the judgment debtor for the nonpayment of the amount due.

judicial foreclosure. N. A type of foreclosure proceeding used in some states that is handled as a civil lawsuit and conducted entirely under the auspices of a court.

jumbo loan. N. A loan that exceeds the limits set by Fannie Mae and Freddie Mac.

junior lien. N. A lien that is subordinate to a senior lien and cannot be satisfied until the senior lien is paid.

junior mortgage. N. A mortgage subordinate to the claim of a prior lien or mortgage. In the case of a foreclosure, a senior mortgage or lien will be paid first.

jurisdiction. N. Geographic or topical area of authority and responsibility for a specific government entity.

just compensation. N. The fair market value of a property, paid to the owner when that property is acquired in an eminent domain action.

keeper. N. The metal latch plate in a door frame into which a doorknob latches fit.

kicker. N. See *equity kicker*.

kick-out clause. N. Sales contract clause that allows a seller to accept a contingent offer and then back out to accept a second and better offer without penalty.

kick plate. N. Metal plate that covers the bottom portion of the room door; protects against water and kick damage.

kiosk. N. An independent stand for selling merchandise.

kitchenette. N. Tiny kitchen area that is often built into the end of another room such as a room in an efficiency apartment.

kitchen triangle. N. Imaginary triangle extending from the sink to the stove to the refrigerator, used as a measurement to maximize the efficiency of a kitchen by reducing the traveling distance between these appliances.

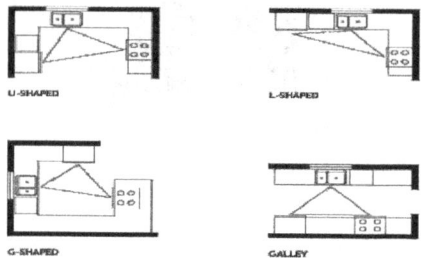

Kitchen triangle

kit home. N. Structure containing prefabricated parts, which is assembled by a contractor.

knee wall. N. A wall-like structure that supports roof rafters.

knob-and-tube wiring. N. An old-fashioned wiring system that has been replaced by fuses and circuit breakers.

labor burden. N. Employer's charges for employee benefits, wages, taxes, insurance, etc.

laches. N. Delay or negligence in asserting one's legal rights, potentially leading to estoppel of the negligent party's suit.

laminated shingles. N. Shingles with a shake-like appearance. Synonymous with *architectural shingles*, *three-dimensional shingles*.

land. N. (1) The surface of the earth. (2) Real estate that is often held for investment purposes.

land contract. N. A type of creative financing in which a down payment is made and periodic payments are made at intervals to pay off the balance. The purchaser may use, occupy, and enjoy the land, but no deed is given by the seller until the total price is paid off.

land cost. N. Total cost of purchasing a land parcel, including purchase price, closing costs, commission, and finance charges.

land development. N. Process of developing raw land by planning and building homes, shopping centers, schools, or churches. Initially, the development process includes construction of streets, sewers, utilities, and other resources.

land economics. N. The branch of economics that concentrates on the economic attributes of land and the economics of agriculture.

land lease. N. Lease that includes only the land and no structures.

land loan. N. Loan used to purchase land. There is more risk connected with the purchase of unimproved land than improved property; therefore, a mortgage for unimproved land will usually have a higher interest rate.

landlocked. ADJ. No access to a public thoroughfare, except through adjacent property.

landlord. N. Person or business that owns property that is rented out to tenants. See also *lessor*.

landlord's lien. N. Landlord's right to receive the value of the tenant's property to pay for unpaid rents or for damages to the lease premises.

landmark. N. A fixed object serving as a boundary mark for a tract of land. Synonymous with *monument*.

land reclamation. N. The process of upgrading unusable land through making physical improvements, such as draining and filling a swamp.

land residual technique. N. An appraisal method of estimating the value of land when given the net operating income and the value of improvements.

land sale-leaseback. N. Sale of land and immediate leasing back by the original owner, whereupon the original owner can realize the capital value of the property and still retain its use.

landscape. N. The area surrounding a home, which consists of grass, plantings, etc.

landscape architect. N. Professional, with a degree, who is trained in the design and planning of horticulture.

landscape contractor. N. One who implements the plans of the landscape architect or designer by doing the planting and upkeep.

landscape designer. N. Professional, without a degree, who is trained in the design and planning of horticulture.

landscape fencing. N. Use of shrubs or trees as a boundary around property.

landscaping. N. The design and planting of trees or other plants on a lot.

land tenure. N. The rights and duties of using and holding property.

land use intensity. N. A measure of the extent to which a land parcel is developed in conformity with zoning ordinances.

land use map. N. An official map indicating intensity of land use in a zoned urban area.

land use planning. N. Activity generally conducted by a local government that provides public and private land use recommendations consistent with community policies, and is generally used to guide decisions on zoning.

land use regulation. N. Government ordinances, codes, and permit requirements intended to make the private use of land and natural resources conform to policy standards.

land use succession. N. Changes in the predominant use of a neighborhood or area over a period of time. Contributing factors to this change include the physical aging of residents leading to the turnover of ownership, business districts expanding into the area, and the physical aging of the building structures.

land value. N. The value of the land in a sale where the total sale price includes land as well as any improvements to the land.

land value map. N. Map delineating property values over a designated area.

lap. V. To cover the surface of one shingle or roll with another.

late charge. N. The penalty a borrower must pay when a payment is made a stated number of days (usually fifteen) after the due date.

latent. N. Something that is hidden or overlooked and may only be realized at a later time.

latent defect. N. Problem that is not obvious, but that may manifest itself at a later point in time, such as lead paint or bad wiring.

late payment. N. Payment received after the due date.

lath. N. A building material of narrow wood, metal, gypsum, or insulating board that is fastened to the frame of a building to act as a base for plaster, shingles, or tiles.

latitude. N. (1) Surveying measure of the angular distance, as measured in degrees, north or south from a fixed point. (2) The loosening of rules and restrictions in certain loan covenants.

lattice. N. An open framework of crisscrossed wood or metal strips that form regular, patterned spaces.

lawful object. N. An object or action that is authorized, approved, and not prohibited by law.

lawn sprinkler system. N. Pipes, valves, sprinkler heads, etc. that are installed in a lawn, above and below ground, to water the grass and shrubs.

laying out. V. Using a plan to determine, prior to actual work, the manner in which pieces of a structure or system relate to each other.

lay of the land. N. Idiomatic expression indicating the desire of an individual to understand new surroundings and all of the new environment's nuances, including its quality and character.

leach field. N. Porous soiled area, through which septic tank leach lines run, emptying out the treated liquid waste forced from the tank, which then percolates down through the soil.

leach lines. N. Lines that carry effluent from the septic system out to the leach fields where, when new sewage is added to the tank, it empties into the area of porous soil.

lead. N. A problem metal in many older dwellings, primarily in the form of lead-based paint and lead plumbing. Exposure to lead has been found to be a health risk.

lead-based paint. N. Paint that is considered a health hazard. If a house was built before 1950, it is likely certain lead-based paint was used. For houses built between 1950 and 1978, there is a lesser chance lead-based paint was used. Lead disclosure regulations can vary from state to state. See also *contingency*.

lease. N. A written agreement between the property owner and a tenant that stipulates the conditions under which the tenant may possess the real estate for a specified period of time and rent.

lease agreement. N. Binding agreement containing the terms of a renter's occupancy.

lease buyout. N. An agreement with a landlord to take care of the tenant's lease obligations in other premises.

leased fee. N. The landlord's ownership interest of a property that is under lease.

leasehold. N. Agreement between the lessee and lessor specifying the lessee's rights to use the leased property for a given time at a specified rental payment. As rental payments are made, rent expense is charged. When the rent is paid in advance, a prepaid rent account is recorded that has to be allocated into expense over the rental period. If the prepayment is for a long-time lease, however, it is recorded as a deferred charge and then amortized. The amortization entry for a long-term lease is to charge rent expense and credit leasehold.

leasehold estate. N. A way of holding title to a property wherein the mortgagor does not actually own the property, but rather has a recorded long-term lease on it.

leasehold improvements. N. Fixtures attached to real estate that are generally acquired or installed by the tenant. Upon expiration of the lease, the tenant can generally remove them, provided such action does not damage the property or conflict with the lease.

leasehold mortgage. N. Mortgage collateralized by a tenant's interest, usually structural improvements, in a lease parcel of property. A leasehold mortgage is subordinate to the landlord's land lease since it is a second lien by order of priority on the property.

leasehold value. N. The value of a tenant's interest in a lease, especially when the rent is below market and the lease has a long remaining term.

lease option. N. Agreement specified in the lease that provides the tenant the option to renew the lease for a given time period upon the expiration of the initial lease. Most lease options include the landlord's right to increase the rent upon renewal.

lease-purchase mortgage loan. N. An alternative financing option that allows low- and moderate-income homebuyers to lease a home from a nonprofit organization with an option to buy. Each month's rent payment consists of principal, interest, taxes, and insurance payments on the first mortgage, plus an extra amount that is earmarked for deposit to a savings account in which money for a down payment will accumulate.

lease-purchase option. N. Lease in which the leaseholder has the option to buy the home after a designated period of time (usually three or five years). Part of each rent payment is put aside toward savings for the purpose of accumulating the down payment and closing costs. Nonprofit organizations may use the lease-purchase option to purchase a home that they then rent to a consumer, or leaseholder.

ledger. N. A book of final entry summarizing all of a company's financial transactions, through offsetting debit and credit accounts.

legacy. N. (1) A gift by will. (2) Something handed down by an ancestor.

legal age. N. The official standard of maturity upon which one is held legally responsible for his or her acts; in most states the age is 18.

legal blemish. N. Problem with a piece of property such as a title claim or zoning violation.

legal description. N. A legally acceptable description of real estate including metes and bounds, government rectangular survey, and lot numbers of a recorded plat. All property deeds have a legal description. Synonymous with *property description*.

legal name. N. The name one has for official purposes, usually the first and last name given at birth, that must be used to sign documents, deeds, or contracts.

legal notice. N. Notification of others using the method required by law, which can include a registered letter, advertisement in a designated newspaper, telegram, or other methods.

legal residence. N. Normally, one's permanent home.

legal title. N. A collection of rights of ownership that are defined or recognized by law or that could be successfully defended in a court of law.

legatee. N. Person who receives a legacy from a will.

lender. N. The bank, mortgage company, or mortgage broker offering the loan.

lender fees. N. Fees charged by the lender to obtain a loan, which typically include loan discount points, loan origination fees, credit report fees, appraisal fees, and assumption fees.

lender participation. See *participation mortgage*.

lending agreement. N. Contract in which a borrower agrees to terms of a loan, which include payment dates, interest rate, total cost of the loan, and late payment fees.

lessee. N. An individual who rents property from another. In the case of real estate, the lessee is also known as the tenant.

lessor. N. One who rents property to another. In the case of real estate, the lessor is also known as the landlord.

let. V. (1) To rent a property to a tenant. (2) To award a contract to the bidder for the property with the best offer.

letter of atornment. N. Letter outlining a tenant's formal agreement to be a tenant of a new landlord.

letter of commitment. N. Official notification to a borrower of the lender's intent to grant a loan; it also specifies the terms of the loan and sets a date for the closing.

letter of credit. N. An arrangement, with specified conditions, in which a bank agrees to substitute its credit for a customer's.

letter of intent. N. A written statement expressing the intention to enter into a contract.

level payment income stream. N. An annuity; a series of equal or nearly equal periodic payments or receipts.

level payment mortgage. N. Mortgage in which each payment made by the borrower is equal, each period, with every payment comprised of principal and interest.

leverage. N. The use of a small amount of cash—a 5% or 10% down payment—to buy a piece of property.

liability. N. A financial obligation, debt, claim, or potential loss.

liability insurance. N. Insurance coverage that offers protection against claims alleging that a property owner's negligence or inappropriate action resulted in bodily injury or property damage to another party.

liable. ADJ. Legally responsible or obligated for something.

libel. N. Written statements, about a person or company, that are unfounded, untrue, malicious, and damaging.

LIBOR. ABBRV. London InterBank Offered Rate.

LIBOR-based adjustable rate mortgage. N. The London InterBank Offered Rate (LIBOR) is based on the interest rate at which major international banks are willing to lend and borrow funds for a specified period of time in the London Interbank market. The LIBOR is similar to the prime-lending rate posted by major U.S. banks. A borrower can select an adjustable rate mortgage that adjusts to the LIBOR at specified periods, usually every six months. This type of ARM typically has a per-adjustment period cap of 1% and is offered with either a 5% or 6% lifetime rate cap.

licensed appraiser. N. An appraiser who meets certain state requirements, but lacks the experience of a certified appraiser.

licensee. N. One who holds a license.

license laws. N. Laws that govern activities, such as in real estate.

lien. N. A legal monetary claim against a property that must be paid off when the property is sold.

lienholder. N. One who benefits from or holds a lien.

lien period. N. Time period in which one may carry out a lien on property.

lien release. N. A written document terminating the terms of a lien, which is normally issued after payment has been made in full.

lien theory state. N. State whose laws give a lien on property to secure debt. In the event of default, lenders may foreclose. See also *title theory state*.

life cap. N. Limit on the amount that a loan rate can change during the term of the mortgage. A mortgage whose interest initially begins at 6% and has a life cap of 7% cannot go over the amount of 13%.

life cycle cost analysis. N. An analysis of a building project's expected operating, maintenance, and replacement costs, calculated by an architect.

life estate. N. A freehold equity in an estate, restricted to the duration of the life of the grantee or other stipulated individual.

life tenant. N. One who is allowed the use of real estate during his or her lifetime or the lifetime of another designated party.

lifetime rate cap. N. The maximum interest rate that may not be exceeded on an adjustable rate loan over the life of the loan.

like-kind exchange. See *tax-free exchange*.

like-kind property. N. Similar property that can be exchanged in a nontaxable transaction.

limited liability. N. The restriction of a person's potential losses to no more than the amount invested.

limited occupancy agreement. N. An agreement that allows for occupancy of a premises for a stated period of time if certain terms are met, and is most often used to allow a prospective buyer to obtain possession under a temporary arrangement, usually prior to closing.

limited partner. N. Member of a partnership whose liability for partnership debts is limited to the amount invested in the partnership.

limited partnership. N. A partnership in which there is at least one partner who is passive and limits liability to the amount invested (limited partner) and at least one partner whose liability extends beyond the monetary investment (general partner). Investment groups of various kinds, including real estate syndicates, use this manner of ownership, where a general partner makes the decisions for the group and is primarily liable for losses.

line of credit. N. The maximum preapproved amount that an individual or business can borrow without filing another application.

lintel. N. A horizontal piece over a door or window that carries the weight of the structure above it.

liquid assets. N. Cash and all other assets that can be converted to cash relatively quickly. Liquid assets can include money in savings and checking accounts, money market accounts, and most certificates of deposit.

liquidate. v. (1) To convert assets into money. (2) To dispose of or get rid of.

liquidated damages. N. Amount agreed upon that one party would pay the other in the event of a breach of a contract.

liquidation. N. Conversion of assets into money; the breaking up and selling of a company's assets for cash distribution to its creditors and then owners. Chapter 7 bankruptcy is a liquidation.

liquidation price. N. Cash value or other consideration that can be received in a forced sale of assets. Liquidation value is typically less than that which could be received from selling assets in the normal course of business.

liquidation value. N. Amount a property would bring under an immediate sale, minus costs of the transaction.

liquidity. N. Ability to obtain close to the true value of an asset by converting it into cash, quickly.

liquidity ratio. See *cash ratio*.

lis pendens. N. (*Latin*) Pending lawsuit; recorded notice of the filing of suit, the outcome of which may affect title to a certain piece of property. Synonymous with *notice of pendency*.

listed property. N. Listed property includes passenger autos and other property used for transportation; property generally used for purposes of entertainment, recreation, or amusement; computers not used exclusively at a regular business establishment; cellular telephones; and, other property to be specified by the Internal Revenue Service. Restrictions apply to the depreciation of listed property.

listing

listing. N. A piece of property placed on the market by a listing agent.

listing agent. N. The sales agent who obtains the right from a seller to handle the marketing of a piece of property.

listing broker. N. The real estate broker who is responsible for the listing of a property and who is to represent the interests of the seller. Brokers are licensed and able to run their own companies. Not all agents are brokers.

listing form. N. The prepared form used to specify terms of the listing agreement.

listing inventories. N. The amount of houses for sale within a given market.

litigation. N. The act or process of carrying on a lawsuit. Alternately, the lawsuit itself.

littoral. N. Land abutting a large body of water, such as the ocean or a lake.

littoral rights. N. Rights concerning property adjoining a large body of water, such as an ocean or a lake, and concerning the ability of the littoral property owner to use the shore and the adjoining water.

live-in partnership. N. Two unrelated people who purchase and live in a home together.

live-work space. N. Dwelling designated to conduct a home-based business.

living unit. N. One single dwelling, condo, apartment, house, etc.

load-bearing wall. See *bearing wall*.

load factor. N. Multiplier to a tenant's useable space of a building's common area, expressed in percentages.

loan. N. A sum of money lent for a specified period of time and repayable with interest.

loan application. N. An itemization of basic financial information presented to a lender by a potential borrower. It is lengthy and requires such information as bank account balances and account numbers, employment data and outstanding debts—including loans—and credit cards with names and addresses of creditors.

loan application fee. N. Fee charged by a lender to cover the initial costs of processing a loan application. The fee may include the cost of obtaining a property appraisal, a credit report, and a lock-in fee. Additional fees will be charged at closing.

loan commitment. See *commitment*.

loan discount point. See *discount point*.

loan officer. N. Official representative of a lending institution who is empowered to loan money within certain guidelines.

loan origination. N. The process by which a mortgage lender brings into existence a mortgage secured by real property.

loan origination fee. N. Commonly referred to as points, the loan origination fee covers the administrative costs of processing the loan. One point is 1% of the mortgage amount. For example, a $100,000 mortgage with a loan origination fee of one point would mean the borrower pays $1,000.

loan rate. N. The interest rate charged for a loan.

loan term. N. The amount of time that is set for the repayment of a mortgage or loan; conforming loans are usually fifteen or thirty years.

loan-to-value percentage. N. The relationship between the principal balance of the mortgage and the appraised value (or sales price if it is lower) of the property. For example, a $100,000 home

loan-to-value ratio

with an $80,000 mortgage has a loan-to-value percentage of 80%. ABBRV. *LTV percentage.*

loan-to-value ratio. N. The ratio of the loan amount to the value of the property. The value is equal to the purchase price or appraised value, whichever is lower. ABBRV. *LTV ratio.*

lock-in. N. A lender's guarantee of an interest rate for a set period of time. The time period is usually that between loan application approval and loan closing. The lock-in protects the borrower against rate increases during that time.

lock-in clause. N. Clause inserted in a loan agreement guaranteeing a quoted interest rate for a specific period of time.

lock-in period. N. Period of time during which a lender guarantees to the buyer a specified interest rate, regardless of a rise in market rates. The longer the time period of the guarantee, the more points charged.

lodging. N. A portion of total support. Lodging includes the fair rental value of a room, apartment, or house in which the dependent lives, a reasonable allowance for the use of furniture and appliances, and all utilities.

log cabin. N. Home constructed of rough-hewn timbers in the style of many early American homes.

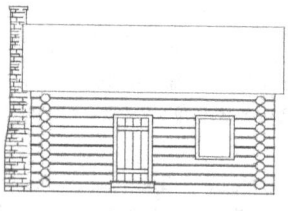

Log cabin

London InterBank Offered Rate. N. The rate charged by banks for very large loans to other banks; this rate is set by a small group of big banks in London. ABBRV. *LIBOR*.

long-term asset. See *fixed asset*.

long-term capital gains and losses. N. Gains and losses on the sale or exchange of capital assets that have been held for more than twelve months. A net long-term capital gain is the excess of long-term gains over long-term losses, or vice versa for a net long-term capital loss.

long-term financing. N. Normally a permanent mortgage, which lasts at least ten years.

long-term lease. N. Lease contract that lasts for at least five years.

long-term liabilities. N. Debts that are payable beyond a one-year period.

louver. N. A vented opening into the home that has a series of horizontal slats and is arranged to permit ventilation but to exclude rain, snow, light, and insects or other living creatures.

low-ball offer. N. A substantially below-market-value offer. Low-ball offers tend to increase in frequency the longer a property stays on the market.

low density. N. A low concentration of housing units in a specific area.

low-documentation loan. N. Mortgage requiring only minimum verification of income and assets.

low down payment loan. N. Home loan requiring only a small down payment to obtain financing for purchase of a home.

low-income housing

low-income housing. N. Housing specifically intended for those people living below a specified income level.

LTV. ABBRV. Loan-to-value ratio.

lumens. N. Unit of measure for total light output; the amount of light falling on a surface of one square foot.

lump sum contract. N. Contract that provides for a set fee for the service to be performed.

MACRS

MACRS. ABBRV. Modified accelerated cost recovery system.

main home. N. Regular, permanent place of residence.

maintenance. N. Periodic expenditure needed to preserve a property's original status rather than to improve the property; activity required to compensate for wear and tear.

maintenance and emergency repairs clause. N. A section of a lease that indicates major and minor repairs and states who pays for what.

maintenance bond. N. Warranty bond given to guarantee that the necessary work will be done by the contractor doing repairs. This type of bond normally has a specific period of time.

maintenance fee. N. Monthly assessment by homeowners' associations on owners and used for maintenance and repair of common areas.

main water shut-off valve. N. Primary valve that halts the flow of water into the home.

maker. N. Any person, company, or legal entity that signs a check or note to borrow money.

mall. N. (1) A public area connecting individual stores in a shopping center. Modern malls are often enclosed, enabling all-weather access. (2) An entire regional shopping center.

mall stores. N. Retail stores in a shopping center other than the anchor tenant, which is normally a larger store.

managed-completion lots. N. Lots in which buyers choose between one of several builders.

management agreement. N. A contract between the owner of a property and the party who agrees to manage it. Fees are usually 4–10% of the rental income.

management fee. N. The cost of professional property management, with a fee typically set at a fixed percentage of total rental income generated by the managed property.

management survey. N. Survey of the maintenance requirements for a commercial or industrial rental property for the purpose of preparing a management agreement.

mansard roof. N. A roof with four sides that slope upward from the roof edge to the square peak.

mantel. N. The facing of stone, marble, or other material around a fireplace.

manufactured housing. N. Homes and dwellings that are not built at the home site and moved to the location. Manufactured housing units must be built on a permanent chassis at a factory and then transported to a permanent site and attached to a foundation. All manufactured homes must be built to meet standards set forth by the U.S. Department of Housing and Urban Development. The standards focus on such aspects as design, strength, energy efficiency, and fire resistance. Manufactured housing represents one of the fastest growing housing markets in the United States. Nearly all of the mortgage products are available for owners of manufactured housing.

manufactured wood. N. A wood product such as a truss or beam that is manufactured out of other wood pieces or mechanically fastened to form a larger piece.

margin. N. The lender's retail markup on the mortgage. If the index rate for an adjustable rate mortgage is 5% but the lender has a 2.5 percentage-point margin, the rate the borrower will pay is 7.5%.

marginal land. N. Property that is difficult to cultivate and has poor income potential.

marginal tax bracket. N. The amount of income tax that an investor would pay on the next dollar of income. Generally, the marginal rate is higher than the average rate because of the progressive tax rate structure.

marginal utility. N. The additional worth or utility received when purchasing an additional unit of a commodity or service identical to the one being purchased. There is often no need for the second item; therefore, its value is marginal.

margin of security. N. Buffer amount between the value of the collateral and the principal balance of the obligation.

marital deduction. N. Tax-free amount transferable to a spouse.

marketable title. N. Title so free from defects that there is no question as to the owner. Any court would enforce this title.

market absorption rate. N. The rate at which a market can absorb additional units of supply without causing market saturation and severe price distortions.

market analysis. N. Research of the supply and demand condition of the real estate market and specific properties in a specific area to discover future trends.

market approach. N. Method of valuing a property through examination and comparison of recent sales of comparable properties.

market area. N. A regional area from which one can expect the greatest demand for a specific product or service.

market comparison approach. N. Method of appraising real estate based on a market comparison of neighboring properties having similar characteristics to ascertain what it could cost to substitute a similar property for the current one.

market conditions. N. Factors that, at a particular point in time, affect the sale or purchase of a home.

market data approach. N. Analysis of real estate sales data to appraise real estate values.

market delineation. N. The process of defining the geographic extent of the demand for a specific property.

market price. N. The actual open market price paid in a transaction where real estate is traded.

market rent. N. Rent that a comparable unit would command if offered in a competitive real estate rental market.

market research. N. Surveys of the area in which a product or service is to be offered, which are done to determine the cost of doing business, any competition, potential sales, etc.

market risk. N. Uncertainty in the value of real estate due to market, economic, political, or other conditions.

market segmentation. N. The process of defining the socioeconomic characteristics of the demand for a specific property.

market study. N. Study of real estate activities including demand, price, locational influence, and current trends.

market value. N. Independently appraised value of real estate in a free competitive market. The highest price a buyer would pay and the lowest price a seller would accept, assuming that both were willing but not compelled to do so.

markup. N. The additional amount added to a bid or price, which includes overhead, profit, excess costs, etc.

masonry. N. The brick or stonework on a building.

masonry wall. N. Wall comprised of brick, stone, cement, etc.

master association. N. A homeowners' association in a large condominium or planned unit development project that is made up of representatives from associations covering specific areas within the project. In effect, it is a "second-level" association that handles matters affecting the entire development, while the "first-level" associations handle matters affecting their particular portions of the project.

master deed. N. Deed filed by the developer or converter of a condominium for the purposes of recording all of the individual condominium units owned within a condominium complex.

master lease. N. A lease in an apartment or office building that controls subleases.

master limited partnership. N. Unincorporated combination of limited partnerships in real estate together as a group. It is generally formed by a roll-up of existing limited partnerships that own property and typically has the advantage of ownership interests that are more marketable than individual limited partnership.

master mortgage loan. N. The mortgage debt existing on a building used for cooperative housing. While each co-op tenant-shareholder is obligated for a portion of the loan, this debt is separate from the loans that may have been used to purchase the individual co-op shares.

master plan. N. Document that describes, in narrative and with maps, an overall development concept including both present property uses as well as future land development plans. The plan may be prepared by a local government to guide private and public development or by a developer on a specific project.

master planned community. N. Development built according to a plan that includes commercial buildings, educational facilities, homes, and community facilities.

mast head. See *entrance cap*.

mastic. N. A pasty material used as a cement (as for setting tile) or a protective coating (as for thermal insulation or waterproofing).

material defect. N. Problem in a specific property that could affect the property's value or salability.

material fact. N. Information about a piece of property that could affect its salability and might change an individual's decision to purchase.

material participation. N. A tax term introduced by the 1986 Tax Reform Act, defined as year-round active involvement in the operations of a business activity on a regular, continuous, and substantial basis.

material participation income. N. Active income, as distinguished from passive income, from employment as well as business and other for-profit activities in which the taxpayer takes a significant and active role.

maturity. N. The date on which the principal balance of a loan, bond, or other financial instrument becomes due and payable.

maximum financing. N. A mortgage amount that is within 5% of the highest loan-to-value (LTV) percentage allowed for a specific

mechanical systems

product. Thus, maximum financing on a fixed-rate mortgage would be 90% or higher, because 95% is the maximum allowable LTV percentage for that product.

mechanical systems. N. A home's plumbing, wiring, heating, and cooling systems.

mechanic's lien. N. An encumbrance filed against a property by subcontractors or suppliers to seek payment for labor, services, or supplies provided to improve the property.

median. N. (1) Strip of land that separates the lanes of opposing traffic. (2) The midpoint in a range of numbers.

median price. N. House price that falls in the middle of the pricing of the total number of homes for sale in a specific area.

mediate. V. To settle a matter by conciliation with use of a neutral third party.

mediation. N. A method of resolving disputes in which a neutral party tries to resolve contract differences.

Megan's Law. N. A federal law requiring states to develop programs for communities when convicted sex offenders are released and move into their neighborhoods. This data is maintained on a public register in that community.

merged credit report. N. A credit report that obtains information from the three big credit reporting companies: Equifax, Experian, and Trans Union Corp.

merger of title. N. The forming of two or more parcels of property under one title.

metes and bounds. N. A land surveying method of describing land in terms of shape and boundary dimensions.

metropolitan statistical area. N. One or more counties having a population of at least fifty thousand. ABBRV. *MSA*.

MGIC. ABBRV. Mortgage Guaranty Insurance Corporation.

millwork. N. Moldings, jambs, etc. that are formed into their finished shapes by removal of excess material.

mineral rights. N. Ownership rights to the minerals or other precious resources in one's property. The privilege of gaining income from the sale of oil, gas, and other valuable resources found on land.

minimum down payment. N. The smallest amount of money that a purchaser is allowed to provide toward the purchase price of a house under a lender's guidelines for a mortgage. Down payments on residential property in the United States have typically been 20% of the purchase price.

minimum lot area. N. The smallest lot area required or allowed for building under the municipal zoning code.

minimum rated. N. Risk in insurance, charging the lowest rate accorded an insurance policy covering a minimum risk classification situation.

mini-warehouse. N. A building separated into relatively small lockable individual units, typically with a garage door–type opening that provides storage.

minority discount. N. A reduction from the market value of an asset because the minority interest owner(s) cannot direct the business operations.

minority interest. N. Ownership of less than 50% of an entity.

mint condition. N. A house or anything that is as close to new as possible.

misrepresentation. N. An untrue statement, whether unintentional or deliberate. Misrepresentation is a form of fraud that could lead to cancellation of a contract or other liability.

mission house. N. A nineteenth-century style of housing that resembles the old mission churches and houses of Southern California. It has a tile roof, arch-shaped windows and doors, stucco walls, and pyramid roof.

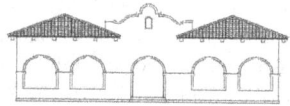

Mission house

miter joint. N. The joint of two pieces at an angle that bisects the joining angle.

mixed-income housing. N. Area of houses at widely varying prices.

mixed-use commercial project. N. Commercial building having several different uses blending together, such as retail shops on the first floor, professional offices on upper floors, and a restaurant on the top floor.

mixed-use development. N. Combination development of several different functions within one area such as residential space combined with a commercial establishment.

MLS. ABBRV. Multiple Listing Service.

model furnishings. N. Interior furnishing included in a model unit, which is chosen to highlight the features of the model unit to show it to its best advantage.

mobile home. N. Premanufactured structure, often constructed of metal, that is designed to be transported to a site and semipermanently attached.

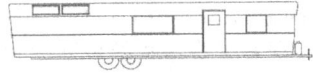

Mobile home

mobile home park. N. Site where mobile homes are located in a subdivision of plots designed for these homes as mandated by municipal zoning laws. They provide necessary utilities and often include recreational facilities.

model unit. N. A representative home, apartment, or office space used as part of a sales campaign to demonstrate the design, structure, and appearance of units in a development.

modernize. N. To upgrade a facility by installing up-to-date equipment, making contemporary cosmetic improvements, and deleting obsolete facilities.

modification. N. Change in the terms of an agreement.

modified accelerated cost recovery system. N. The method of depreciation introduced by the Tax Reform Act of 1986. The modified accelerated cost recovery system (MACRS) is not an entirely new system of depreciation, but rather a series of significant modifications to the accelerated cost recovery system. MACRS is mandatory for most depreciable assets placed in service after December 31, 1986, and was available on an optional basis for assets placed in service after July 31, 1986, and before January 1,

1987. Under MACRS, costs of qualified property are written off over predetermined periods. Examples of property classes include twenty-year property, residential property, and nonresidential property. Twenty-year property includes property such as farm buildings. Residential rental property is comprised of rental buildings or structures (including mobile homes) for which 80% or more of the gross rental income is derived from dwelling units. This class excludes hotels and motels, and is depreciated over 27.5 years. Nonresidential real property includes real property that is not residential rental property; this property is depreciated over 31.5 years. ABBRV. *MACRS*.

modified annual percentage rate. N. An index of loan cost based on the standard annual percentage rate, but adjusted for the time the borrower expects to hold the loan.

modified gross lease. N. A real estate lease that requires the tenant to pay the base rent and requires the landlord to cover all costs related to property operations.

modify. V. To change or alter, slightly or partially.

modular housing. N. Dwelling units constructed from components prefabricated in a factory and erected on the site.

module. N. Unit of standardized size or design, which can be arranged or fitted together with other modules in a variety of ways.

molding. N. Decorative trim around windows and door openings, ceilings and floors, etc., used to give a better appearance as well as to provide protection from jagged edges and to help in preventing drafts. Molding may be made from any material, but the most often used material is wood. Synonymous with *moulding*.

money market account. N. A savings account that provides bank depositors with many of the advantages of a money market fund. Certain regulatory restrictions apply to the withdrawal of funds from a money market account.

money market fund. N. A mutual fund that allows individuals to participate in managed investments in short-term debt securities, such as certificates of deposit and Treasury bills.

Monterey architecture. N. A nineteenth-century style, two-story house with a balcony across the front at the second-floor level, which was adopted from the early California-Spanish period.

Monterey architecture

monthly association dues. N. Monthly payment paid to a homeowners' association and used for maintenance and repair in housing that has communal areas.

monthly fixed installment. N. That portion of the total monthly payment that is applied toward principal and interest. When a mortgage negatively amortizes, the monthly fixed installment does not include any amount for principal reduction.

monthly payment mortgage. N. A mortgage that requires payments to reduce the debt once a month. A monthly mortgage payment is composed of principal, interest, taxes, and insurance.

month-to-month tenancy. N. A tenancy in which no written lease is involved, rent being paid monthly. It can be renewed for each succeeding month or terminated at the option of either party with sufficient notice.

monument. See *landmark*.

mortgage. N. A legal document by which real property is pledged as security for the repayment of a loan. The items stated in the mortgage include the homeowner's responsibility to pay principal, interest, taxes, and insurance on time; pay to maintain hazard insurance on the property; and, adequately maintain the property. The mortgage also includes fulfilling the obligations found in the mortgage note. Should a borrower consistently fail to meet these requirements, a lender can seek full repayment of the balance of the loan, foreclose on the property, or sell the property and use the proceeds to pay off the loan balance and foreclosure costs. A deed of trust is used instead of a mortgage in some states. Synonymous with *home loan*.

mortgage acceleration clause. N. Provision in a mortgage that gives the lender the right to demand repayment of the entire loan under certain circumstances, such as default, property sale, change of title, or refinance.

mortgage amortization. N. Repayment of a loan on a scheduled installment basis. As a loan is amortized, the equity in the associated property is increased. In the early years, the bulk of each payment goes toward interest rather than principal.

mortgage-backed security. N. Certificates that pass through principal and interest payments to investors.

mortgage banker. N. Company that uses its own money to provide home loans and then usually sells them to institutional investors—like insurance companies—and Fannie Mae.

mortgage banking company. N. Company that originates and services mortgages, i.e makes loans to consumers. Mortgage companies then typically sell these loans to other lenders and investors. Some mortgage companies may be subsidiaries of

mortgage bond. N. Bond collateralized by real estate. Two kinds of mortgage bonds are senior mortgages, which have a first claim on assets and earnings, and junior mortgages, which have a subordinate lien. A mortgage bond may have a closed-end provision that prevents the firm from issuing additional bonds of the same priority against the same property, or may be an open-end mortgage that allows the issuance of additional bonds having equal status with the original issue.

mortgage broker. N. An individual or company that brings borrowers and lenders together for the purpose of loan origination. Mortgage brokers typically require a fee or commission for their services. Mortgage brokers presently originate more than half of the residential loans in the United States.

mortgage commitment. N. A written document stating the willingness of a lender to give a mortgage to a mortgagor. The commitment will provide a time period in which the mortgage will be given and an indication of the interest rate to be charged. The mortgage will be granted at closing of title.

mortgage constant. N. Ratio of annual mortgage payments divided by the initial principal of the mortgage. Applies only to loans involving constant payment.

mortgage correspondent. N. One who services mortgage loans for a fee.

mortgage credit certificate. N. Qualified taxpayers who receive a mortgage credit certificate from a state or local government to buy, rehabilitate, or improve their main homes may claim a credit for a percentage of their home mortgage interest. The percentage is set by the government and ranges from 10% to 50%. If the percentage exceeds 20%, the maximum credit is $2,000 per year. The itemized deduction for home mortgage interest must be reduced by the amount of the credit. The credit is not refundable, but any portion that is unused because it exceeds tax liability may be carried over to the following three years where it can be added to any credit for the current year. The credit is computed on Form 8396. Mortgage credit certificates may be subject to a recapture rule if the home is sold within nine years.

mortgage discount. N. One-time charge assessed by a bank or other financial institution at the closing of buying real estate. One discount point translates to 1% of the initial mortgage amount.

mortgagee. N. The lender in a mortgage agreement.

Mortgage Guaranty Insurance Company. N. Private company established in 1957 in Milwaukee, Wisconsin, to provide private mortgage insurance (PMI) to mortgage lenders granting mortgages to mortgagors not having at least a 20% down payment upon application. The Mortgage Guarantee Insurance Company indemnifies the mortgage lending company should the mortgagor go into foreclosure because of a default. The cost of PMI is included in the closing costs by the mortgagee. ABBRV. *MGIC*.

mortgage instrument. N. Written mortgage document that states the terms of the mortgage including the interest rate, length of payments, payment dates, and remedies the bank is entitled to in the event of the mortgagor's failure to pay as required, including late charges.

mortgage insurance. N. Insurance on some loans, which protects lenders from possible default by the borrower. Conventional loans with down payments of less than 20% of the home value usually require private mortgage insurance. Federal Housing Association and Veterans Administration loans have different insurance guidelines.

mortgage insurance premium. N. The amount paid by a mortgagor for mortgage insurance, either to a government agency such as the Federal Housing Administration or to a private mortgage insurance company. ABBRV. *MIP*.

mortgage interest deduction. N. Tax write-off allowed by the Internal Revenue Service, where owners may deduct annual interest payments made on real estate loans.

mortgage lien. N. A lien that secures the loan which funded the purchase of that property.

mortgage life insurance. N. Specific insurance that will pay off a mortgage if the borrower dies while the debt is still outstanding. Similar in purpose to private mortgage insurance in that both insurance products secure repayment to the lender if the borrower dies.

mortgage loan. N. A loan for which real estate serves as collateral to provide for repayment in case of default.

mortgage market. N. The interest rate and terms competing mortgage lenders are offering to potential mortgagees.

mortgage note. N. Legal document obligating a borrower to repay a loan at a stated interest rate during a specified period of time. The agreement is secured by a mortgage or deed of trust or other security instrument.

mortgage origination fee. N. A charge for work involved in preparing and servicing a mortgage application.

mortgage out. N. The obtaining of financing at or in excess of the construction or acquisition cost of a project. The acquirer, or developer, is not required to invest any equity capital.

mortgage participation agreement. N. A written agreement between institutional investors to buy or sell ownership shares in mortgages.

mortgage payment table. N. Table used to compute the monthly mortgage payment that consists of principal repayment and interest. A type of loan amortization formula is used. The tables have monthly payments for any combination of loan size, interest rate, and term.

mortgage pool. N. Collection of loans of similar nature that are sold as a unit in the secondary market or used to back a security, which is then sold in the capital markets.

mortgage REIT. N. Type of real estate investment trust (REIT) that does not own property, but gives construction or permanent mortgage loans for major projects.

mortgage-related closing costs. N. Costs generally associated with a loan application. They vary, but some of the most common ones include loan origination fees, loan discount points, credit report fees, assumption fees, prepaid interests, and escrow accounts.

mortgage release price. N. Amount required to pay off the full balance of the mortgage at a given time. This amount is the principal balance plus any prepayment penalty.

mortgage relief. N. Acquired freedom from mortgage debt, generally through assumption of the mortgage by another party or debt retirement.

mortgage requirement. N. The amount of a periodic payment, including interest and principal, required for a mortgage payment.

mortgage risk rating. N. The amount of risk for a mortgagee in granting a mortgage loan. A principle in mortgage risk is that a maximum of 28% of the mortgagor's salary should be devoted to the mortgage payment and 33% should be put toward total debt payments (including the mortgage).

mortgage servicing. N. Monitoring and administering a mortgage loan after it has been made. This may include collecting monthly payments, record keeping, and dealing with foreclosures.

mortgagor. N. The borrower in a mortgage agreement.

moulding. See *molding*.

move-in condition. N. House that is ready for an occupant.

move-up buyer. N. Owner of one home who is looking to buy a bigger, more expensive home.

moving expenses. N. An adjustment to income permitted to employees and self-employed individuals who move for work-related reasons, providing certain requirements are met. IRS Form 3903 is used to compute deductible moving expenses.

MSA. ABBRV. Metropolitan statistical area.

mullion windows. N. Two side-by-side windows, separated by a very small distance.

multidwelling property. N. Residential property containing individual units for several households within the same structure.

multidwelling units. N. Properties that provide separate housing units for more than one family, although they secure only a single mortgage.

multifamily mortgage. N. A residential mortgage on a building that is designed to house more than four families, such as a high-rise apartment complex.

multiple dwelling. N. More than one dwelling unit sharing a common wall and roof.

multiple listing. N. Listing agreement used by a broker who is part of a multiple-listing organization.

Multiple Listing Service

Multiple Listing Service. N. Service combining the listings of all the available homes, except those sold by the owner, in a specific area into one database. ABBRV. *MLS*.

multiple offers. N. More than one offer to purchase a property, which usually occurs in a seller's market.

Multi-ply Construction. N. Type of construction that uses more than one layer of plywood in a structure, which increases the fire rating.

municipal. ADJ. Having to do with a city or town, or the city or town's government.

Municipal Housing Inspector. N. Inspector employed by a city or county to verify that all contractors are meeting building codes on all construction sites within a specific area.

municipal sewer. N. The main sewer system to which private sewers are connected. Synonymous with *public sewer*.

muniments of title. N. Documentation of ownership, such as a deed.

mutual consent. N. Two or more parties agreeing to something.

mutual funds. N. A trust or corporation formed to invest the funds it obtains from shareholders in diversified securities.

mutual savings banks. N. State-chartered banks, which are owned by the depositors and operated for their benefit. Many of these banks hold a large portion of their assets in home mortgage loans.

mutual water company. N. Business entity providing water services in a particular locality.

NAA

NAA. ABBRV. National Apartment Association.

NAEBA. ABBRV. National Association of Exclusive Buyers Agents.

NAHB. ABBRV. National Association of Home Builders.

NAHI. ABBRV. National Association of Home Inspectors.

NAIFA. ABBRV. National Association of Independent Fee Appraisers.

nail pops. N. Nails in load-bearing parts of new homes that pop out slightly because of settling of the structure.

NAR. ABBRV. National Association of Realtors®.

NAR Code of Ethics. N. Formal code of ethics and standards of practice by which members of the National Association of Realtors® must abide.

NAREIT. ABBRV. National Association of Real Estate Investment Trusts.

National Apartment Association. N. This group consists of sixty state and local associations of managers, investors, developers, owners, and builders of apartment houses and other residential rental property. ABBRV. *NAA*.

National Association of Exclusive Buyers Agents. N. National organization of buyer's brokers whose members do not accept property listings. ABBRV. *NAEBA*.

National Association of Home Builders. N. Founded in 1942 and located in Washington, D.C., with 155,000 members and 824 local groups, this organization of homebuilders provides educational, political information, and research services. Its publications include *Builder Magazine, Forecast of Housing Activity,*

Housing Economics, *Housing Market Statistic*, and the *Nation's Building News*. Its membership consists of single, multifamily, and commercial builders.

National Association of Home Inspectors. N. Professional association of independent home inspectors who meet the group's education and performance requirements. ABBRV. *NAHI*.

National Association of Independent Fee Appraisers. N. This organization of real estate appraisers offers the professional designations of IFA for members, IFAS for senior members, and IFAC for appraiser-counselors. ABBRV. *NAIFA*.

National Association of Real Estate Investment Trusts. N. A trade association that serves real estate investment trusts (REITs), it collects data on the performance of REITs and prepares definitions to be used. ABBRV. *NAREIT*.

National Association of Realtors®. N. Trade organization for real estate agents and brokers who become members by agreeing to abide by the organization's code of ethics. ABBRV. *NAR*.

National Association of Review Appraisers & Mortgage Underwriters. N. This organization awards the Certified Review Appraiser designation.

National Council of Real Estate Investment Fiduciaries. N. Organization that collects historical data on various institutional-grade property types, sorted by geographic areas. Publishes data on income and value changes. Its index, called the Russell/NCREIF Real Estate Performance Report, is often cited as the benchmark for institutional real estate performance. ABBRV. *NCREIF*.

National Council of State Housing Agencies. N. Nonprofit clearinghouse of information on local and state housing agencies.

National Electric Code®

National Electric Code®. N. An affiliate of the National Fire Protection Association that sets a minimum standard for electrical installations.

National Flood Insurance Program. N. Provides coverage for those people suffering real property losses as a result of floods. Any real estate located in a flood plain area cannot be financed through a federally regulated lender unless flood insurance is purchased. ABBRV. *NFIP*.

National Society of Real Estate Appraisers. N. An affiliate of the National Association of Real Estate Brokers, its purpose is to formulate rules of ethics and professional conduct and enforce these rules to the benefit of its members. Designations given by this organization after the completion of a course of study and final exam are Master Real Estate Appraiser, Certified Real Estate Appraiser, and Residential Appraiser. ABBRV. *NSREA*.

national tenant. N. A well-known and more substantial lessee with a presence in most of the United States.

natural vacancy rate. N. The average vacancy rate for a rental property market that would result if supply and demand were in balance. This level to which vacancy rates adjust over the long term is a benchmark by which current vacancy rates in the market are considered high or low.

NCREIF. ABBRV. National Council of Real Estate Investment Fiduciaries.

necessary. N. An expense that is appropriate and helpful in furthering the taxpayer's business or income-producing activity. See also *ordinary and necessary business expenses*.

needs-based pricing. N. Asking price based on the amount of funds required to pay off the seller's mortgage, the cost of remodeling, or the purchase of another house.

negative amortization. N. When the outstanding balance of a loan grows larger because each monthly payment is too small to cover both the principal and interest of that loan. This sometimes happens with adjustable rate mortgages.

negative cash flow. N. When operating expenses exceed income.

negative net worth. See *deficit net worth*.

negative-slope driveway. N. A driveway that drops from street level to the garage.

negligence. N. A lack of such reasonable care and caution as would be expected of a prudent person. A penalty may be assessed if any part of an underpayment of tax is due to negligent or intentional disregard of rules and regulations.

negotiable. ADJ. Able to be changed through discussions and modifications. N. Something that is legally transferred to another by endorsement or proper delivery.

negotiable instrument. N. A promise to pay money, transferable from one person to another.

negotiate. V. To bargain or attempt to reach an agreement through discussion.

negotiation. N. The process of bargaining that precedes an agreement.

neighborhood. N. A district or locality characterized by similar or compatible land uses, often with a major street for shopping or restaurants.

neighborhood life-cycle. N. A generalized pattern that describes the physical and social changes that residential areas experience over time. This life cycle includes the phases of birth, early growth, maturity, and decline. Neighborhoods decline because of several reasons, including the physical aging and deterioration of the building structures, as well as the aging of the population. Architectural obsolescence makes these neighborhoods less attractive and the intrusion of business or industrial areas detract from the overall quality.

neo-traditional planning. N. Community planning that focuses on new-home development with grid-street patterns, prominent front porches, backyard garages, multi-use buildings, and housing clustered near commercial service areas.

net. N. (1) The amount remaining after certain adjustments have been made for debts, deductions, or expenses. (2) The proceeds from the sale of an investment minus the purchase price, including commissions and other expenses. (3) Term used on an invoice to indicate that the full amount is payable.

net assets. See *net worth*.

net book value. See *depreciated cost*.

net capital. N. A firm's net worth, minus deductions taken for any assets that might not easily be converted into cash at their full value.

net cash flow. N. The income that remains for an investment property after the monthly operating income is reduced by the monthly housing expense, which includes principal, interest, taxes, and insurance for the mortgage, homeowners' association dues, leasehold payments, and subordinate financing payments.

net current assets. See *working capital*.

net earnings. See *net profit*.

net floor area. N. Usable floor area after deducting stairs, walls, and similar features.

net income. N. The amount remaining when expenses are deducted from gross income.

net income multiplier. N. The price of an asset divided by the net income it generates in a given period of time—for rental property, usually one month.

net interest margin. N. The dollar difference between interest income and interest expenses, usually expressed as a percentage of average earning assets.

net investment. N. The level of investment minus equipment depreciation.

net leasable area. N. Floor space in a building that is actually under lease and able to be rented to tenants. Nonleasable areas include hallways, building foyers, and other areas devoted to utilities, elevators, etc. Synonymous with *net rentable area*. ABBRV. *NLA*.

net lease. N. Lease in which the lessee pays not only a fixed rental charge, but also expenses on the rented property, including maintenance. Synonymous with *triple net lease*.

net listing. N. Listing agreement where the broker's commission is an amount above a net price set by the owner. If that price is not met, a commission is not earned.

net long-term gain. N. In taxation, the excess of total long-term gains minus total long-term losses on the sale of real estate. Long-term classification is for real estate held one year or more and is reported on Schedule D of IRS Form 1040 (for sole proprietors) or Form 1120 (for corporations).

net loss. N. The excess of total expenses over rental revenue for a real estate business.

net operating income. N. Income from property or business after operating expenses have been deducted, but before deducting income taxes and financing expenses (interest and principal payments). ABBRV. *NOI*.

net operating loss. N. A net loss for the year attributable to business or casualty losses. In order to mitigate the effect of the annual accounting period concept, the law allows taxpayers to use an excess loss of one year as a deduction for certain past or future years. In this regard, a carryback period of two years (three or five years for certain losses) and a carryforward period of twenty years is allowed. ABBRV. *NOL*.

net present value. N. Method of determining whether expected performance of a proposed investment promises to be adequate. The difference between the present value of cash inflows generated by real estate and the amount of the initial investment. The present value of future cash flows is computed using the cost of capital (minimum desired rate of return, or hurdle rate) as the discount rate. ABBRV. *NPV*.

net proceeds. N. Amounts received from the sale or disposal of real property less all relevant deductions, i.e., direct costs associated with the sale or disposal.

net profit. N. The amount of money earned after all expenses have been deducted from the total revenue. Synonymous with *bottom line*, *net earnings*.

net realizable value. N. Expected selling price of property minus costs to sell. Net amount received upon sale. Gross receivables less allowance for doubtful accounts, representing the expected collectibility of those receivables.

net rentable area. See *net leaseable area*.

net tangible assets. N. Net worth minus goodwill.

net worth. N. Total worth of a person or entity once the liabilities are deducted from the assets. Synonymous with *capital net worth, net assets*. See also *individual net worth, corporation net worth, deficit net worth*.

net yield. N. The return on an investment after subtracting all expenses.

neutral wire. N. A color-coded wire carrying electricity from an outlet back to the service panel.

New England colonial. N. Early American style, two-and-a-half-story, box-like house that is usually symmetrical, square, or rectangular with side or rear wings. The traditional material is narrow clapboard siding. The roof is usually the gable type covered with shingles. It is often referred to as a Saltbox Colonial.

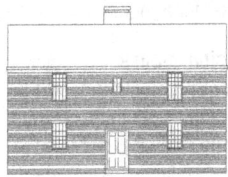

New England colonial

New England farmhouse. N. A simple, box-shaped house with clapboard siding and a gable roof.

new town. N. A large, mixed-use development designed to provide residences, general shopping, services, recreation, and employment. Typically near a metropolitan location, it can enjoy the associated amenities.

NFIP. ABBRV. National Flood Insurance Program.

niche. N. A small recessed area in a wall, traditionally arched at the top.

NIMBY. ABBRV. Not in my backyard.

NLA. ABBRV. Net leasable area.

NNN. ABBRV. Triple net lease.

no bid. N. Decision by the Veterans Administration to pay the guarantee amount to the lender instead of acquiring the property in foreclosure when a loan it has guaranteed goes into default. The result is that often the lender obtains the property at sale.

no cash-out refinance. N. A situation in which a new mortgage will cover the remaining balance of the first loan, closing costs, or any liens, but does not yield more than 1% of the principal in cash.

no competition lots. N. A lot in which the buyer's home will be constructed by a particular builder.

no deal, no commission clause. N. A clause placed in a listing agreement stating that no commission will be paid to the broker until the property title has actually been transferred. If not in the listing agreement, commission is payable once a ready, willing, and able buyer, who agrees to the terms of the sale, is found.

no documentation loan. N. Loan application where verification of income is not required and is typically granted in cases of large down payments.

nodular cast iron. N. Commonly used for fittings, valves, pipe, etc., this iron has magnesium or cerium added while in a liquefied state so that it can be formed into globular nodules. It has good corrosion-resistant characteristic, and is less brittle than gray cast iron. Synonymous with *ductile cast iron*.

NOI. ABBRV. Net operating income.

NOL. ABBRV. Net operating loss.

nominal loan rate. N. The loan rate stated on the face of the loan note, which is different from the effective interest rate. If points are charged, the effective rate will be higher.

no money down. N. General term referring to real estate acquisition strategies based on seller-provided financing and/or existing loan assumption and minimal use of cash-down payments; using a maximum amount of leverage to achieve maximum profits from real estate investments.

nonalienation clause. N. Clause in a document forbidding an individual from selling or transferring a subject property to another and is normally used in a trust where the grantor of the trust believes the designated beneficiary to be a spendthrift.

nonassumption clause. N. A statement in a mortgage contract forbidding the assumption of the mortgage by another borrower without the prior approval of the lender.

nonbusiness bad debt. N. A bad debt loss not incurred in connection with a creditor's trade or business. A nonbusiness bad debt is deductible as a short-term capital loss and is allowed only in the year the debt becomes entirely worthless.

noncash expense. N. An income statement expense for which no cash was spent, such as amortization or depreciation.

nonconforming loan. N. Any loan that does not meet the qualifications or is too large to be purchased by Freddie Mac or Fannie Mae.

nonconforming use. N. Property use that is in violation of the current zoning ordinance, but had been in use prior to the zoning ordinance's enactment.

noncurrent asset. N. An asset that is not easily convertible to cash or not expected to become cash within the next year, such as fixed assets, leasehold improvements, and intangible assets.

nondestructive examination. N. Any type of inspection for quality or condition that does not damage the object in question, such as visual inspections, X-rays, and ultrasounds.

nondisturbance clause. N. (1) An agreement in mortgage contracts on income-producing property that provides for the continuation of leases in the event of loan foreclosure. (2) A clause in a sales contract where the seller retains mineral rights, which provides that exploration of minerals will not interfere with surface development.

nonexclusive listing. See *open listing*.

nonjudicial foreclosure sale. N. Foreclosure sale enabled in those states permitting the use of a power of sale clause to be inserted into a mortgage or deed of trust empowering the mortgagee to advertise and sell a property at a foreclosure sale upon the mortgagor's payment default. A nonjudicial foreclosure sale enables a foreclosure action without a formal judicial action.

nonliquid asset. N. An asset that cannot easily be converted into cash.

nonloadbearing wall. N. A wall that does not support the weight of a building or structure. A nonloadbearing wall can be removed without affecting a building's structural soundness.

nonperformance. N. Failure or refusal to perform a specified action; the failure to fulfill contractually agreed-upon terms or actions. Nonperformance creates a liability, which can enable a judicial damage action.

nonrecourse. N. No personal liability. Lenders may take the property pledged as collateral to satisfy a debt but have no ability to take the other assets of the borrower.

nonrecovery property. N. Property that does not qualify for a cost recovery deduction under either the accelerated cost recovery system (ACRS) or the modified accelerated cost recovery system (MACRS), or property the taxpayer elects to exclude from the ACRS or MACRS by choosing a depreciation method not based on a number of years.

nonrecurring closing costs. N. Costs that are one-time-only fees for such items as appraisals, loan points, credit reports, title insurance, and home inspections.

non sequitur. N. (*Latin*) It does not follow; the illogical conclusion of a statement or phrase, in which it does not relate to anything previously said.

nontaxable exchange. N. An exchange on which no gain or loss is recognized in the current year.

nontaxable income. N. Income that is by law exempt from tax.

normal wear and tear. N. Physical depreciation arising from age and ordinary use of the property.

notarize. V. To attest, in one's capacity as a Notary Public, to the genuineness of a signature.

Notary Public. N. Public officer given the right to authenticate a document, accept a person's oath, administer depositions, and to conduct other activities in commercial business. An official seal is used by the Notary.

note. N. A legal document that obligates a borrower to repay a mortgage loan at a stated interest rate during a specified period of time. A note represents a borrower's promise to pay the lender according to the agreed-upon terms of the loan, including when and where to send the payment. The note lists any penalties that will be assessed if payments are not made on time, and also warns the borrower that the lender can call the loan, i.e., demand repayment of the entire loan before the end of the term, if terms are violated. Synonymous with *promissory note*.

note rate. N. Interest rate specified in a mortgage note.

notes payable. N. Funds due to a lender.

notes receivable. N. Funds owed to an individual or entity by a borrower.

notice. N. Official communication of a legal action or one's intent to take an action.

notice of cancellation clause. N. (1) Notice, often in writing, in which an individual or business gives a notice of termination to another, pursuant to a cancellation provision in a contract to forestall future liability. (2) A notice given between an insurer and a re-insurer, or an insurer and an insured, of the termination of a contract or policy at the time of renewal, or in the latter case, for nonpayment of premium payments.

notice of cessation. N. Notice to one or more individuals to cease and desist from performing a particular action.

notice of default. N. Written notice to a borrower that a default has occurred and that legal action may be taken.

notice of nonresponsibility. N. Clearly stated notice that an owner or operator will not assume responsibility for an inherent risk.

notice of pendency. See *lis pendens*.

notice to pay rent or quit. N. Notice to a tenant by a landlord to either pay rent due or vacate the premises. With a long-term lease, the notice may list penalties. Synonymous with *three-day notice to quit*.

notice to quit. N. Notice to a tenant to vacate rented property.

not in my backyard. N. The response sometimes given by neighborhoods and communities to proposed changes or development. ABBRV. *NIMBY*.

notorious possession. N. Open and active occupancy of a piece of real estate that acknowledges the fact that the borrower is in possession. Notorious possession is one of the important tests when finding for or against a claim of property under adverse possession, which is to acquire land by unauthorized but lengthy occupation.

novation. N. Substitution of a subsequent borrower in place of the original borrower, who is then released from liability. This must be done with the approval of the lender.

NPV. ABBRV. Net present value.

NSREA. ABBRV. National Society of Real Estate Appraisers.

nuisance. N. (1) A land use whose associated activities are incompatible with surrounding land uses. (2) An activity by a property owner that annoys or seriously disturbs other property owners, making it discomforting to use their own property.

nuncupative will. N. An oral will made by a testator before a sufficient number of witnesses, just prior to death. Nuncupative wills depend on the oral testimony of those witnesses present as proof. They are illegal in several states and are enforceable in others only if they meet specific guidelines.

obiter dictum. N. (*Latin*) Something said in passing; opinion of a judge that has no direct legal or binding effect on the outcome of a pending judicial decision.

obligatory. N. The legal requirement of a debtor, or obligor, to pay a debt and the legal right of a creditor, or obligee, to demand satisfaction of a debt or enforce payment in the event of default.

obligee. N. The person to whom a debt or obligation is owed.

obligor. N. The person responsible for paying a debt or obligation.

occupancy. N. Residing in or using real estate.

occupancy cost. N. Charges to the tenant pursuant to his or her lease including rent, operating expense increases, parking charges, moving expenses, remodeling costs, etc.

occupancy date. N. The date the buyer is able to legally move into the purchased or leased property. A buyer should add a provision to his or her purchase offer that holds the seller responsible for paying rent should he or she not move out on or prior to the agreed-upon date. This provides the buyer money to pay for his or her own unexpected housing or lodging expenses if the property is not ready after closing.

occupancy level. N. The number of units currently occupied in a facility, neighborhood, or city, stated as a percentage of total capacity.

occupancy ratio. N. The ratio of rented or leased space to the total amount of space available.

Occupational Safety and Health Administration. N. This government agency oversees safety in most places of work. ABBRV. *OSHA*.

off-balance-sheet financing. N. Financing from sources other than debt and equity offerings, such as joint ventures, research and development partnerships, and operating leases.

offer. N. An expression of willingness to purchase a property at a specified price; presenting for acceptance a price for a property parcel; the bid price in a real estate or security transaction. Once a prospective buyer has made an offer, the seller has the opportunity to accept, decline, or make a counteroffer. If the offer is accepted, the buyer will receive a ratified sales contract. This contract is the starting point for working with an approved lender to get a mortgage if the buyer is not already preapproved.

offer and acceptance. N. Two requirements of a contract forming mutual consent combined with valuable consideration, which are the major elements of a contract.

offeree. N. One who receives an offer.

offerer. N. One who extends an offer to another.

offer to purchase. N. A proposal to buy property at a specified price whereupon the seller has the options of accepting or rejecting the offer or making a counteroffer.

office building. N. A structure primarily used for the conduct of business, such as administration, clerical services, and consultation with clients and associates. Such buildings can be large or small and may house one or more business concerns.

office condominium. N. Condominium with units that are used as commercial offices. The purchaser of an office condominium owns the title to the individual office unit and not to the property. Maintenance fees are assessed to each owner.

Office of Interstate Land Sales Registration. N. Division of the U.S. Department of Housing and Urban Development (HUD) that regulates offerings of land for sale across state lines. ABBRV. *OILSR*.

Office of Thrift Supervision. N. Federal agency created by the Financial Institutions Reform, Recovery and Enforcement Act of 1989 (FIRREA) to regulate and supervise federally chartered savings and loan associations. The Office of Thrift Supervision takes over the thrift regulatory duties exercised by the Federal Home Loan Bank Board prior to passage of FIRREA. This agency is part of the U.S. Department of the Treasury. ABBRV. *OTS*.

office park. N. A planned development specially designed for office buildings and supportive facilities, such as restaurants. Some office parks are designed to attract specific tenants such as medical services.

offset statement. N. An occupant's expressed interest in property that is being rented.

off-site. N. Not on a construction site proper; in a different location. For example, utility lines are brought into a development from off-site.

off-site costs. N. Expenditures related to construction but not directly on the property in question.

off-site improvements. N. The portions of a subdivision or development that are not directly on the lots to be sold but enhance them in some way.

off the books. N. Payments for which no formal record is kept.

OILSR. ABBRV. Office of Interstate Land Sales Registration.

omnibus clause. N. (1) Provision in a will that stipulates that any assets not enumerated still pass to the heirs. (2) A clause in liability insurance policies that extends coverage to unnamed others beyond the insured.

one-time charge. N. An expense that a company recognizes in a single reporting period, and that the company claims is unlikely to recur in the future.

one-year adjustable rate mortgage. N. This adjustable rate mortgage (ARM) offers a low initial interest rate that adjusts annually after the first year. The rate cap per annual adjustment is usually 2%; the lifetime adjustment caps can be 5% or 6%. This type of mortgage may be good for the buyer who anticipates a rapid increase in income over the first few years of the mortgage. It lets the buyer maximize his or her purchase power immediately, and it is also good for homeowners who do not plan to live in a home for more than a few years. One-year ARMs come in terms from ten to thirty years. The most typical ones are ten, fifteen, or thirty years. The one-year ARM is most often indexed to the weekly average yield of U.S. Treasury securities adjusted to a constant maturity of one year. Can be used to buy one-family principal residences, including condos, and planned unit developments. Manufactured homes are also eligible, but they must be built on a permanent chassis at a factory and then transported to a permanent site and attached to a foundation. An advantage to one-year ARMs is that some let borrowers convert to a fixed-rate loan at certain adjustment intervals. Generally, conversions to fixed-rate mortgages are allowed at the third, fourth, or fifth interest rate adjustment dates.

ongoing costs. N. Homebuyers should not forget that there are ongoing costs associated with owning a home. They include, but are not limited to: monthly mortgage payments; mortgage insurance; homeowners' insurance; property taxes; and, utilities such as gas, oil, water, and electricity. Homebuyers should also consider the costs of maintaining their home. These costs include everything from cleaning and minor repairs to yard work and painting. Condominium owners and people living in planned unit developments should factor in any homeowners' association fees or similar costs.

online real estate listings. N. Properties listed for sale on the Internet.

on-site improvements. N. Direct enhancements made to the physical nature of a property such as renovating a building, installing a new driveway or parking lot, and gardening.

on-site management. N. The direct management of property at its location, where its functions may include showing prospective tenants the facilities, collecting rents, and doing upkeep on the property.

open-end lease. N. A lease contract providing for a final additional payment on the return of the property to the lessor, adjusted for any value change.

open-end loan. N. A loan where the borrower may add to the principal without renegotiating the terms of the loan. Additional sums borrowed under the terms of an open-end loan will have the same rate of interest and life of loan terms as the original loan. A home equity loan is often open ended.

open-end mortgage. N. A mortgage or trust deed that can be increased by the mortgagee (borrower). The mortgagee may secure additional money from the mortgagor (lender) through an agreement that typically stipulates a maximum amount that can be borrowed.

open hole inspection. N. When an engineer inspects an open excavation and examines the earth to determine the type of foundation (caisson, footer, wall on ground, etc.) that should be installed in the hole.

open house. N. Method of showing a home that is for sale. The house is left open for prospective buyers to see, and may be advertised in the newspaper or outside the house.

open housing. N. See *fair housing*.

opening escrow. N. The deposit given by a buyer of property is delivered to the escrow agent, who retains it for the seller.

open listing. N. Property marketed by a number of brokers simultaneously. This type of agreement permits a real estate agent to sell the property while allowing the homeowner of other agents to attempt to make the sale. Synonymous with *nonexclusive listing*.

open listing agreement. N. Listing agreement given to many brokers and agencies. The property owner pays a commission only to the broker who actually produces a buyer for the property.

open mortgage. N. (1) Mortgage that has matured or is overdue and, therefore, open to foreclosure at any time. (2) A mortgage that does not have a prepayment clause and permits the mortgagor to repay the mortgage at any time without paying a penalty.

open occupancy. See *fair housing*.

open space. N. Land that is left undeveloped for use as parks, walking paths, etc., by those living in a planned community.

open space ratio. N. The ratio in a development of open space to developed land.

open year. N. A taxable year for which the statute of limitations has not yet expired.

operating asset. N. Asset that contributes to the regular income from a company's operations.

operating costs. N. The day-to-day expenses incurred in running a business, such as sales and administration, as opposed to production. Synonymous with *operating expenses*.

operating expense ratio. N. Mathematical equation obtained by dividing operating expenses by gross income.

operating expenses. See *operating costs*.

operating income. N. The day-to-day earnings of a company, before deduction of interest payments and income taxes. Synonymous with *earnings before interest and taxes*.

operating lease. N. Regular rental of property between the lessee and lessor for a fee. An operating lease does not satisfy the criteria for a capital lease.

operating leverage. N. Automatic increases in the net operating income or cash flow of income-producing real estate when income and expenses increase at the same rate; further enhanced when expenses are fixed.

operating statements. N. Financial reports on the cash flow of a property.

opinion of title. N. A certification, usually from an attorney, as to the validity of title to property being sold, after examining the abstract of title. The opinion of title is essential to obtaining title insurance or a mortgage, and to the transfer of title.

opportunity cost. N. A term used in economics; when taking a particular action, the loss of the value of the next best action.

optimize. V. To increase the efficiency or effectiveness of a process as much as possible.

option. N. An agreement to buy or sell property on or before a specified date at an established price. The sale or exchange of an option to buy or sell property results in capital gain or loss if the property is a capital asset.

optionee. N. One who receives or purchases an option.

option listing. N. Listing agreement that also gives the listing broker the right to purchase the property.

option listing agreement. N. A contract given for a consideration where an optionor gives an option to the optionee for the right, but not the obligation, to purchase property within a certain period of time, at a certain price. If the option is not exercised within the specified period of time, it will expire.

optionor. N. Person or business that gives or sells an option.

option to purchase. N. Contract that gives one the right, without any obligation, to purchase a property within a certain period of time and at a certain price, subject to conditions.

oral agreement. N. Arrangement that is not memorialized in writing and is not usually legally binding.

oral contract. N. Contract not in writing. Some oral contracts are enforceable but those applicable to the sale of real estate are unenforceable.

ordinance. N. Law enacted by local authorities to govern the activities of people or things, such as land usage.

ordinary. N. Common and accepted in the general industry or type of activity in which the taxpayer is engaged. It is one of the tests for the deductibility of expenses incurred or paid in connection with a trade or business; for the production of income; for the management, conservation, or maintenance of property held for the production of income; or, in connection with the determination, collection, or refund of any tax.

ordinary and necessary business expenses. N. Tax term describing current and necessary business expenses, which are allowed as deductions. Ordinary and necessary business expenses do not include long-term capital losses.

ordinary annuity. N. A series of equal payments, with each payment occurring at the end of each equally spaced period.

ordinary income. N. Defined by the Internal Revenue Code to include salaries, fees, commissions, interest, dividends, and many other items. Taxed at regular tax rates, contrasted with long-term capital gains, which receive more favorable tax treatments.

ordinary interest. N. Interest based on a 360-day year instead of a 365-day year. The former is referred to as "simple interest" and the latter is termed "exact interest." The difference between the two types of interest can be significant when a substantial investment is involved.

ordinary loss. N. For income tax purposes, a loss that is deductible against ordinary income. Usually more beneficial to a taxpayer than a capital loss, which has limitations on deductibility.

original cost. N. The total costs associated with the purchase of an asset, for accounting purposes.

original equity. N. The amount of cash initially invested by the underlying real estate owner.

original principal balance. N. The total amount of principal owed on a mortgage before any payments are made.

origination fee. N. A fee paid to a lender for processing a loan application. The origination fee is stated in the form of points. One point is 1% of the mortgage amount.

origination process. N. The process by which a loan is funded, including the due diligence process financial structure and lender committee approvals.

OSHA. ABBRV. Occupational Safety and Health Administration.

other people's money. N. The use of borrowed funds by people or businesses to increase the return on an investment. The term implies that debt can be used to maximize investment profits or minimize the risk of personal loss.

OTS. ABBRV. Office of Thrift Supervision.

outbuilding. N. Any structure located on a lot in addition to the house or main building. Outbuildings can be barns, shops, sheds, etc.

outgo. N. Money paid out.

outlay. N. An expenditure.

outstanding balance. N. The amount still unpaid and owed on a debt, loan, or other financing agreement.

overage. N. (1) In leases for retail sales, amounts to be paid, based on gross sales, over the base rent. (2) A selling price received for property in excess of the expected price. (3) An excessive amount; surplus.

overage income. N. Rental based on a percentage of sales or profit that is in addition to the constant rental amount.

overall rate of capitalization. N. The percentage relationship of net operating income divided by the purchase price of property; net operating income (NOI) of property relative to its market value. If rental income property worth $1,000,000 results in a NOI of $100,000, the overall return is 10%. Synonymous with *overall rate of return*.

overall rate of return. See *overall rate of capitalization*.

overbuilding. N. A situation in a given area where there has been more real estate construction than the market can absorb within a reasonable time.

overhang. N. Outward projecting eave-soffit area of a roof; the part of the roof that hangs out or over the outside wall. Synonymous with *cornice*.

overhead. N. The ongoing administrative expenses of a business, such as rent, utilities, and insurance.

overhead ratio. N. Operating expenses divided by the sum of taxable equivalent net interest income and other operating income.

overimproved property. N. Property whose sale price is not high enough to recoup the costs of its improvements

overimprovement. N. Land improvement that is more extensive than the surrounding neighborhood justifies or that can be economically warranted.

override. N. Percentage of a commission or a fee paid to someone higher in the organization or above a certain amount.

overriding royalty. N. Percentage of royalties derived from an oil and gas lease payable to someone other than the property lessor. It is a net royalty interest in the oil and gas recovered at the surface, free of all operating expenses.

owner. N. The person or entity to whom a piece of property belongs. In real estate, the person or entity with title to the property.

owner financing. N. A property purchase transaction in which the property seller provides all or part of the financing.

owner occupant. N. A tenant of a residence who also owns the property.

owner of record. N. The person or persons who, according to the public records, own a particular property.

owner's equity. See *net worth*.

ownership. N. Title or a legal right to possess something; the state of possessing something.

ownership form. N. Method of owning real estate. Ownership form has important consequences for income tax, estate tax, corporate income tax, and survivorship. Real estate may be owned by one or more persons. Forms of ownership include tenancy in common, joint tenancy, a tenancy by the entirety, tenancy in severalty, partnership, limited partnership, and corporation.

ownership in severalty. N. Ownership of property by one person or one legal entity (corporate ownership).

ownership rights to realty. N. The right to possess, exclusively occupy, enjoy, control, and dispose of real estate. Ownership rights to realty are granted by the ownership of a title to real property.

package mortgage. N. A single mortgage on both the purchased real estate and the durable personal property in the house.

packaging. N. (1) Transfer of both real estate and personal property. (2) The putting together of a group of properties to be sold together, possibly at a discount price because several items are sold in combination.

pad site. N. An individual freestanding site for a retailer, often adjacent to a larger shopping center.

PAM. ABBRV. Pledged account mortgage.

panel. N. Part of a wall, ceiling, or a flat piece of building material that covers the part of the surface on which it rests.

paneling. N. Thin strips or sections of wood or wood material applied to a wall.

paper profit. N. An increase in value above original cost or basis that would be realized if the property were sold. Until a sale occurs, the increased value is not recognized in the accounts. When the property is sold, there will be a realized gain or loss. Synonymous with *book profit*.

paper title. N. Passing of title to property that is in fact not valid.

parapet. N. Protective, low wall along a roof or any edge or below a terrace.

parcel. N. A piece of land that is usually part of a larger acreage.

parking strip. N. The strip of grass between the sidewalk and the street in front of a house.

parol. N. Statement made verbally.

parol evidence. N. Oral evidence, rather than that contained in documents.

parol evidence rule. N. Permits oral evidence to augment a written contract in certain cases.

partial eviction. N. Removal of a tenant from a portion of a rented or leased premise.

partial interest. N. Ownership of a part of the ownership rights to a parcel of real estate.

partial payment. N. A payment that is not sufficient to cover the scheduled monthly payment on a mortgage loan.

partial release. N. Release of a portion of a property from a mortgage.

partial taking. N. Purchase of part of a piece of property or property rights when condemnation takes place. The owner must be justly reimbursed.

participation loan. See *equity kicker*.

participation mortgage. N. Agreement between a mortgagee and a mortgagor that allows the lender to have a percentage of ownership in that particular property. Allows the lender to share in part of the income or resale proceeds. Synonymous with *lender participation*.

partition. N. (1) The division of real estate between owners, giving each an undivided interest. (2) An interior wall dividing an area into two or more rooms or separate areas.

partition action. N. Court action to order a compulsory sale of real estate owned jointly between two or more owners. A partition action divides the proceeds of a real estate sale among the joint owners rather than physically dividing the real estate into separate undivided interests.

partnership. N. An agreement between two or more entities to go into business or invest. Either partner may bind the other, within the scope of the partnership. Each partner is liable for all the partnership's debts. A partnership normally pays no taxes, but merely files an information return. Various items of partnership income, expenses, gains, and losses flow through to the individual partners and are reported on their personal income tax returns.

party wall. N. Boundary wall between two properties, built along the line separating the properties, partly on each parcel of land. Either owner has the right to use the wall and has an easement over that part of the adjoining owner's land covered by the wall.

passive income. N. Generally, income from rents, royalties, dividends, interest, and gains from the sale of securities. A meaning created by the Tax Reform Act of 1986 distinguishes passive income or loss from active income and portfolio income.

passive income generator. N. A business or investment that produces passive income that can be used to offset passive losses. ABBRV. *PIG*.

passive investor. N. Someone who invests money but does not manage the business or property.

passive loss. N. Tax term referring to a loss from a passive activity, such as ownership by nonoperation of rental real estate.

passive solar system. N. A system that supplies solar heat without the use of electric fans or pumps.

pass-through certificates or securities. N. Securities supported by a pool of mortgages. The principal and interest are due monthly on the mortgages and are passed through to the investors who bought the pool.

patent. N. Exclusive right granted by the government to a company or person to use, manufacture, and sell a product or process for a seventeen-year period without interference or infringement by other parties.

patent defect. N. Visible deficiency in a piece of property such as a sagging porch, etc.

patio. N. An interior courtyard or a paved backyard area.

payback period. N. The amount of time required for cumulative estimated future income from an investment to equal the amount initially invested. It is used to compare alternative investment opportunities.

payee. N. One who receives a payment, such as by cash, check, money order, or promissory note.

payer. N. One who makes a payment.

payment bonds. N. Security that a contractor's bills will be paid from the money given by the client, so that the client is not held liable.

payment cap. N. Consumer safeguards that limit the amount monthly payments on an adjustable rate mortgage may change. Since they do not limit the amount of interest the lender is earning, they may cause negative amortization.

payment change date. N. The date when a new monthly payment amount takes effect on an adjustable rate mortgage or a graduated-payment adjustable rate mortgage. Generally, the payment change date occurs in the month immediately after the adjustment date.

penalty. N. Money that will be paid by a person or business for violating a statute or legal court order. Also may be assessed for violating the provisions of a contract.

penalty clause. N. Provision in a contract that specifies the dollar amount or rate an individual to the contract must pay for not conforming to its terms.

pendente lite. N. (*Latin*) Pending the suit; a lawsuit in which the outcome is pending.

penthouse. N. A luxury housing unit located on the top floor of a high-rise building.

percentage lease. N. Where the rent is based on a percentage of the sales volume made inside the leased premises.

percolation test. N. A test used to determine the ability of soil to accommodate a septic system.

per diem interest. N. Interest charged or accrued daily.

perennial. N. Any plant that produces leaves, flowers, and seeds from year to year, such as irises or peonies.

perfecting title. N. Removing a cloud or claim from the title.

pergola. N. An arbor with an open roof of rafters supported by posts or columns.

periodic payment cap. N. For an adjustable rate mortgage, a limit on the amount that payments can increase or decrease during any one-adjustment period.

permanent mortgage. N. Mortgage for an extended period of time, usually ten to twenty-five years.

permits. N. Documents that provide legal permission to undertake a project and are usually given by local government agencies; often required for major home improvement projects. Some of the most common permits are for general projects or permits that require property owners and builders to meet specific local building codes. Check with local government to determine if there are building restrictions in historic areas or in environmentally sensitive areas.

per se. ADV. (*Latin*) By itself. A per se matter is one that is alone and not connected to another matter.

personal assets. N. Personal property and other assets a person has in his or her estate.

personal financial statement. N. Document showing the financial health of an individual that may be requested for a loan application.

personal liability. N. An individual's responsibility for a debt.

personal residence. N. The dwelling unit that one claims as one's primary home. This dwelling establishes one's legal residence for voting, tax, and legal purposes. A residence is not limited to a house. Condominiums, cooperative apartments, townhouses, mobile homes, and houseboats can all qualify as residences. Synonymous with *principal residence*.

personal-use property. N. Property owned for personal well-being and enjoyment, which includes a taxpayer's home, vehicles, furniture, clothing, and other property.

per stirpes. N. (*Latin*) A legal way to distribute the estate to include the descendents of a deceased litigee.

per-unit allocation. N. Allocating common or central costs to each unit of property.

pest-control inspection. N. Professional inspection to determine whether or not there are insects in a dwelling; usually required by a lender.

phased building. N. Portions of construction are completed prior to other portions being started. For example, the exterior of a building would be completed prior to interior work being started. Construction is often done in this manner to protect the incomplete parts and for economic reasons.

physical damage insurance. N. Insurance coverage for any risk that can cause physical damage to the insured item.

physical depreciation. N. Decline in value of property due to all causes of age and action of the elements.

pier. N. A rectangular masonry support column.

piggyback loan. N. (1) Loan, with participation by two or more lenders, in the financing of a single mortgage. (2) A combination of a construction loan with a permanent loan commitment.

pipefitter. N. Contractor whose job is to install piping for steam, cooling, hot water, etc.

piping area drawing. N. Drawing done by the layout person for the piping system, which shows, to scale, the routing of the piping system using either elevation views or plan and section views.

piping codes and standards. N. Local and state laws establish codes applicable to piping and piping systems. Materials are covered by American Society for Testing and Materials standards with other regulations being set by the American Society of Mechanical Engineers, American Petroleum Institute, American Water Works Association, American National Standards Institute, American Welding Society, Manufacturers Standardization Society of the Valve and Fitting Industry, Plastics Pipe Institute, Copper Development Association, and the Uni-Bell PVC Pipe Association.

piping isometric drawing. N. Three-dimensional drawing that shows the layout, sizes, and dimensions of the piping system of a structure.

PITI. ABBRV. Principal, interest, taxes, and insurance.

PITI reserves. N. A cash amount that a borrower must have on hand after making a down payment and paying all closing costs for the purchase of a home. The principal, interest, taxes, and insurance (PITI) reserves must equal the amount that the borrower would have to pay for PITI for a predefined number of months.

plaintiff. N. In a legal action, the party initiating the suit to obtain relief from the court against the defendant.

PLAM. ABBRV. Price level adjusted mortgage.

planned community. N. Description of a neighborhood built with certain guidelines in mind.

planned unit development. N. A project or subdivision that includes common property that is owned and maintained by a homeowners' association for the benefit and use of the individual unit owners. ABBRV. *PUD*.

planning commission. N. Governmental body that has the responsibility of planning the future development of a jurisdictional area. A planning commission is responsible for developing and managing a zoning ordinance as well as interfacing with a professional planning department.

planning grid. N. Grid that shows the dimensions of a structure in multiples of four to give the builder some choice in locating openings and allowing the matching of vertical and horizontal surfaces. Material lists are matched to the grid, so that use of the standard-sized materials can minimize waste.

plan view. N. A floor plan of a structure that is shown from a horizontal plane above the structure.

plaster. N. A hand-formed wall finish.

plat. N. Plan or map of a specific land area, showing the boundaries of individual properties.

plat book. N. Public record of maps showing the division of streets, blocks, and lots, and providing the measurements of the individual parcels of land.

platform framing. See *western framing*.

plat map. N. Map showing land, with township, streets, improvements, lot lines, etc., within a specific area.

pleadings. N. Formal allegations by all the parties to an action including complaints, answers, and replies to counterclaims.

pledged account mortgage. N. Type of home purchase loan under which a sum of cash contributed by the owner is set aside in an account pledged to the lender. The account is drawn down during the initial years of the loan to supplement periodic mortgage payments. The effect is to reduce the payment amounts in early years. ABBRV. *PAM*.

pledgee. N. Individual to whom a mortgage or property is pledged.

pledgor. N. Individual who is responsible for making the payments on a mortgage on property that has been pledged.

plot. N. Parcel of land or small lot.

plot plan. See *survey*.

plottage. N. The combining of several smaller contiguous land lots to make a larger, more useful and valuable piece.

plottage value. N. The result of combining two or more parcels of land so that the single, larger parcel has more value than the sum of the individual parcels.

PMI. ABBRV. Private mortgage insurance.

pocket door. N. A sliding door that retreats into the wall when opened.

pocket license card. N. Pocket-sized card required for salespersons and brokers in most states. Issued by the state licensing agency, it identifies its holder as a licensee and must be carried at all times business is conducted.

point. N. A one-time charge by the lender for originating a loan; one point is 1% of the amount of the mortgage. See also *loan origination fee*.

policy. N. (1) A real estate owner's rules regarding the use of a property by a tenant; a principal plan or course of action. (2) A written contract in which one party guarantees to insure another against a specified loss.

population density. N. Population per square mile of a given area.

porch. N. A simple covered platform at the entrance to a home or a fully enclosed room on the outside of a residence.

porte cochere. N. A porch-like roof extending over a driveway.

portfolio. N. A group of investment assets.

portfolio diversification. N. Choosing alternative investment instruments having different risk-return features. Diversification provides a lower but acceptable overall return.

portfolio income. N. Income from investments such as interest, dividends, royalties (unless earned in the ordinary course of business), and gains from the sale of property. Portfolio income cannot be used to offset passive activity losses. Real estate investments produce rental and lease payment income. Investments in mortgages and other long-term debt instruments produce interest income, while equity investments generate dividends.

portfolio lender. N. A lender that makes loans with its own funds and keeps the loans on the company's books—i.e., inside the institution's portfolio—rather than selling the loan on the secondary market.

portico. N. A porch supported by a row of columns.

positive cash flow. N. Periodic amounts available to an equity investor after deducting all periodic cash payments from rental income.

positive leverage. N. Profitably using borrowed funds to increase the return on an investment; when the return on the borrowed funds exceeds the after-tax interest costs.

possession. N. The holding, control, or custody of property for one's use, either as an owner or a person with another legal right.

possession by adverse possession. N. Way to acquire title to real estate when an occupant has been in actual, open, exclusive, and continuous occupancy of property for an extended period of time.

possessory action. N. Litigation undertaken to obtain or maintain possession of real property.

potential gross income. N. The amount of income that could be produced by the sale or rental of property or rendering of services.

potentially responsible parties. N. Referring to a superfund site, all owners, operators, transporters, and disposers of hazardous waste. ABBRV. *PRP*.

potential rental income. N. The total of all rents under the terms of each lease, assuming the property is 100% occupied.

power of attorney. N. A legal document that authorizes another person to act on one's behalf. A power of attorney can grant complete authority or can be limited to certain acts or certain periods of time.

power of sale. N. A provision in a mortgage agreement that grants the lender or trustee the right to sell the property upon certain default. The property is to be sold at auction but court authority is unnecessary.

prairie house. N. An early twentieth-century architectural style featuring a long, low roofline with a continuous row of windows and a plain exterior. It is a very open design with long horizontal lines rather than small, secluded rooms. Architectural development is credited to Frank Lloyd Wright.

Prairie house

preamble. N. Clause at the beginning of a legislative statute explaining its purpose; it neither confers nor increases powers contained within the statute and is, therefore, not an essential element of it.

preapproval. N. The process of determining how much money a prospective homebuyer or refinancer will be eligible to borrow prior to application for a loan. A preapproval includes a preliminary screening of a borrower's credit history. Information submitted during preapproval is subject to verification at application. Confirmation of the amount to be borrowed by a person whose ability to borrow has already been assessed by the lender. Being preapproved for a mortgage can make a buyer more attractive to a seller.

preapproval letter. N. A letter from a lender confirming the amount that can be borrowed by a person whose ability to borrow has already been assessed by the lender.

preapproved mortgage. N. A commitment from a lender to provide a mortgage loan to a borrower on stated terms before the borrower has found a property to buy. Real estate agents encourage this type of mortgage because it allows them to make a firm offer when they find something.

prearranged refinancing agreement. N. A formal or informal arrangement between a lender and a borrower wherein the lender agrees to offer special terms (such as a reduction in the costs) for a future refinancing of a mortgage being originated as an inducement for the borrower to enter into the original mortgage transaction.

preclosing. N. Prior to actual closing, all information is available in order to ensure that the appropriate parties properly execute all documents. A preclosing is used primarily when the closing is expected to be complicated by many extraneous factors.

predepreciation profit. N. Profit before considering noncash expenses.

preemptive right. N. Right of a current stockholder to maintain the percentage ownership in a real estate company by purchasing new shares on a proportionate basis before they are issued to the public. It allows existing stockholders to keep the value and control they presently enjoy. The new shares may be issued to the current stockholders at a lower price than the going market price. Furthermore, brokerage commissions do no have to be paid.

preexisting use. N. Property use that is in violation of the current zoning ordinance, but had been in use prior to the zoning ordinance's enactment.

prefabricated. ADJ. Constructed in a factory, usually in modules or units, and then assembled where it is to be used.

prefabricated house. N. House made using standardized components that are preassembled on an assembly line in a factory and then erected and assembled on the site. Normally, the prefabricated house is trucked onto the home site where it is installed on a completed foundation.

preforeclosure sale. N. A procedure in which the investor allows a mortgagor to avoid foreclosure by selling the property for less than the amount that is owed to the investor.

prelease. N. The obtaining of lease commitments in a building or complex prior to its being available for occupancy.

premises. N. Land and any existing buildings that are part of a conveyance as noted in a deed.

premium. N. (1) Amount paid for real estate over and above the expected prevailing price; the value of a mortgage or bond in excess of its face amount. (2) A periodic fee paid for insurance protection.

prepaid assets. See *prepaid expenses*.

prepaid expenses. N. Assets that come up on a balance sheet as a result of payments made by a company for goods and services to be received in the future. Synonymous with *prepaid assets*.

prepaid interest. N. Interest that is paid in advance of when it is due. Typically charged to a borrower at closing to cover interest on the loan between the closing date and the first payment date.

prepayment. N. Any amount paid to reduce the principal balance of a loan before the due date; payment in full on a mortgage that may result from a sale of the property, the owner's decision to pay off the loan in full, or a foreclosure. In each case, prepayment means payment occurs before the loan has been fully amortized.

prepayment clause. N. Clause in a mortgage that gives a mortgagor (borrower) the privilege of paying the mortgage indebtedness before it becomes due. Often, there is a penalty for prepayment.

prepayment penalty. N. Fee charged by a lender for a loan paid off in advance of the contractual due date.

prepayment privilege. N. The right of a borrower to retire a loan before maturity.

prepayment terms. N. Provision of credit that applies when a loan is paid.

prequalification. N. The process of determining how much money a prospective homebuyer will be eligible to borrow before he or she applies for a loan.

presale. N. Sale of proposed properties before construction is completed.

prescription. N. Method of obtaining title to property through adverse possession, such as open, notorious, and continuous use of the property for a statutorily prescribed period of time.

prescriptive easement. N. Easement obtained by long-term usage of a property without objection by the owner.

present value analysis. N. Way of valuing real estate that computes the discounted present value of an expected stream of income, including rental income and future capital gains or losses.

present value of annuity. N. The current value of a level stream of income to be received each period for a finite number of periods.

present value of one. N. The value today of an amount to be received in the future, based on a compound interest rate.

present value tables. N. Precalculated tables providing the present values of $1 or an annuity of $1 for different time periods and at different discount rates.

preservation district. N. Zoning designation covering a sensitive environmental area, parkland, scenic area, or historic district, and placing especially strict limitations on private landowners' and nonowners' freedom to change the essential character of sites within the district.

presold home. N. A home that is sold prior to being built.

pressure relief valve. N. A safety vent that relieves excess pressure in a water heater.

pre-tax income. N. The amount earned from a business or investment before deducting income taxes.

preterit. N. Past action(s) of a property owner or tenant.

preventive maintenance. N. Maintenance procedures conducted to prevent later repairs and extend a useful life.

price. See *value*.

price appreciation. N. Increase in the value of real estate or personal property. The price may increase because of a number of factors, such as shortage in supply, improved economy, favorable political environment, tax incentives, increased profitability.

price fixing. N. Illegal effort by competing businesses to maintain a uniform price, such as a commission rate on the sale of real estate.

price level adjusted mortgage. N. Loan in which the interest rate is changed periodically based on the change in a general price index to take into account inflation, such as a yearly adjustment. ABBRV. *PLAM*.

price range. N. Upper and lower limits of what a buyer will pay for a home.

primary beneficiary. N. The person who will receive the benefits of a trust or insurance policy when distribution is made.

primary financing. N. Loan that comes before all other loans in the event of default.

primary lease. N. Rental agreement between the owner and a tenant. When the tenant rents to someone else, it is called a sublease.

primary location. N. Real estate that is located in the best area for its designated use.

primary market area. N. A regional area from which one can expect the greatest demand for a specific product or service.

primary mortgage market. N. Mortgage market in which original loans are made by lenders. The market is made up of lenders who supply funds directly to borrowers and hold the mortgage until the debt is paid.

primary metropolitan statistical area. N. A classification of the U.S. Census Bureau applied to cities within a metropolitan area with a population of one million or more, which would qualify as metropolitan areas on their own, yet are linked to other cities in close proximity. These individual areas are called primary metropolitan statistical areas, while the metropolitan area containing them is called a Consolidated Metropolitan Statistical Area. ABBRV. *PMSA*.

prime contractor. N. Contractor who assumes responsibility for completing a construction project, under contract to the owner, and hires, supervises, and pays all subcontractors.

primer. N. The initial coat of paint that is applied before the final topcoat.

prime rate. N. The interest rate that banks charge to their preferred customers. Changes in the prime rate influence changes in other rates, including mortgage interest rates.

prime tenant. N. Major tenant in an office building or shopping center. The prime tenant occupies more space than the others and will attract customers to the site. They are normally more creditworthy.

principal. N. The amount borrowed and remaining unpaid. It is also the part of the monthly mortgage payment that reduces the remaining balance of a mortgage.

principal balance. N. The outstanding balance of principal on a mortgage. The principal balance does not include interest or any other charges.

principal broker. N. The licensed broker responsible for the operations conducted by the firm.

principal, interest, taxes, and insurance. N. Payment amount calculated to include the principal, interest, taxes, and insurance on an amortizing loan and represents the borrower's actual monthly mortgage-related expenses. ABBRV. *PITI*.

principal payments. N. Payments received on the contract price.

principal place of business. N. The main place where work is performed or business is transacted. Taxpayers who engage in more than one business can have more than one principal place of business. For purposes of the home-office deduction, a principal place of business may also be an area of a taxpayer's home that is used for the management and record keeping portions of the business, provided there is no other fixed location where the taxpayer performs such functions.

principal residence. See *personal residence*.

principle of conformity. N. The concept that a house will be more likely to appreciate in value if it is similar to other houses in the neighborhood.

principle of progression. N. Appraisal term stating that the value of lower-end real estate is enhanced by the proximity of higher-end properties.

principle of regression. N. Appraisal term stating that the value of higher-end real estate can be brought down by its proximity to lower-end properties.

privacy fence. N. Structure erected between two pieces of property.

private mortgage insurance. N. Insurance required on some loans to protect lenders from possible default by borrower. Conventional loans with down payments of less than 20% of the home value usually require private mortgage insurance. ABBRV. *PMI*.

private offering. N. An offering of securities, stock, or debt directly to investors (usually large institutional investors) rather than through the public exchange markets. An advantage of a private placement to a real estate business is that the securities do not have to be registered with the Securities and Exchange Commission.

private property. N. Privately owned property.

private sewer. N. Sewer that belongs to the building where it is installed but discharges into the public sewer. Synonymous with *building sewer*.

privity. N. A mutual interest in the same property or rights established by law or legalized by contract.

probate court. N. Court that has the responsibility of performing probate of wills and administering estates. In certain states, a probate court can appoint guardians for minor children of an estate.

probate or prove. N. Process of establishing the validity of a will in court. Probate court then administers the will as directed or as authorized by the court to settle financial obligations.

probate sale. N. A real estate sale triggered by the death of the owner, with proceeds to be divided among heirs or creditors.

proceeds of condemnation. N. The value of tenant improvements on a property if an eminent domain action occurs.

processing fee. N. Fee charged by most lenders to pay for gathering the information necessary to process the loan.

process time. N. Time needed for performance of an operation through to its completion.

procurement of supplies and services. N. A lease provision that requires the tenant to purchase supplies and services from the landlord or the landlord's visitors.

procuring cause. N. (1) The cause that results in the attainment of a stated goal. (2) In real estate, it refers to the action of the real estate agent or broker, who through their actions in producing a buyer brought about the sale of a property.

production home. N. Mass-produced homes, i.e., a tract of homes built by one builder.

professional appraiser. N. Expert in real estate who has an education in real estate appraisal as well as significant professional experience. A recognized license may be obtained from the Appraisal Institute. However, no national requirement exists as to who may do an appraisal. When an appraisal is done by a federally insured agency, the appraiser must be licensed by the state.

profit. N. The sum remaining after all costs, direct and indirect, are deducted from the income of a business.

profit and loss statement. See *income statement*.

pro forma statement. N. Financial statement with amounts or other information that are completely or partially assumed. The assumptions supporting the amounts are usually provided. The statement may be prepared in determining the possible financial effects of buying or renting property.

programming. N. A written summation by an architect of a project's design objectives, constraints, and criteria.

progress payments. N. In a construction loan, payments made to a contractor as the various construction stages are completed. The contractor uses progress payments to pay the various subcontractors and suppliers as construction proceeds.

project budget. N. Outline of the construction budget and all costs for land, equipment, financing professional services, etc.

projection period. N. The time duration for estimating future cash flows and the resale proceeds from a proposed real estate investment.

promissory note. See *note*.

property. N. The rights that one individual has in lands or goods to the exclusion of all others. Property rights include exclusive occupancy, possession, use, and the right of disposition. Individuals, groups, organizations, and governments may own property.

property and casualty policy. N. Insurance contract providing coverage for risks primarily associated with negligence and acts of omission associated with third-party injuries or property losses. Property and casualty policies normally exclude losses associated with war, riots, and unreasonable negligence.

property brief. N. Summarization of the attributes and characteristics of the property such as those indicated in the legal records.

property damage liability insurance. N. An insurance policy that promises to pay all the legal obligations of the insured due to negligence in which damage to the property has been caused.

property damage liability losses. N. Losses arising from damage to or destruction of property.

property depreciation insurance. N. Insurance protection for the replacement cost of damaged property. Thus, the accumulated depreciation is not subtracted in determining the amount of reimbursement.

property description. See *legal description*.

property insurance

property insurance. N. Insurance affording protection against losses due to damage to or destruction of property or contents therein. Insurance protects assets and any future income thereon from loss such as that incurred from a fire, etc.

property inventory. N. A listing of all assets a person or business owns, their cost, and their appraised value.

property line. N. Official dividing line between two properties. Legal boundary of property.

property management. N. The operation of a property as a business, including rental, rent collection, maintenance, etc.

property report. N. Required by the Interstate Land Sales Full Disclosure Act for the sale of subdivisions of fifty lots or more, if the subdivisions are not otherwise exempt. Filed with the Department of Housing and Urban Development's Office of Interstate Land Sales Registration.

property residential technique. N. In appraisal, a method for estimating the value of property based on estimated future income and the reversionary value of the building and land.

property tax. N. Tax paid on privately owned properties and based on local tax rates and assessed property values.

property tax deduction. N. U.S. tax code allowing homeowners to deduct the amount they have paid in property taxes.

property under contract. N. Real estate being offered for sale that has received a contract for sale but has not yet gone to a closing.

property value. N. Value of a piece of property based on the amount a buyer will pay at any given time.

proposal. N. Detailed presentation of an offering to perform a job for a specified amount under certain conditions. Often given by a subcontractor to a general contractor.

proposal to lease. N. A response to a request for proposal, stating the landlord's position on the various terms and requirements of the prospective tenant.

proprietary lease. N. Lease in a cooperative apartment building; the lease a corporation provides to the stockholders, which allows them to use a certain apartment unit under the conditions specified.

proprietorship. N. Ownership of a business, including income-producing real estate, by an individual, as contrasted with a partnership or corporation.

prorate. V. To allocate percentages of certain expenses to be paid between the buyer and seller at time of closing.

proration. N. The assignment of agreed-upon percentages of certain expenses associated with a piece of property to the buyer or the seller at the time of closing.

proration of taxes. N. The proportionate division of taxes at the closing between the buyer and the seller.

prospect. N. Potential customer or client.

prospect cards. N. File of prospective real estate customers showing their addresses, telephone numbers, times and dates of last contact, types of properties in which they are interested, and their financial capabilities.

prospectus. N. A printed descriptive statement about a business or investment that is for sale, designed to invite the interest of prospective investors. Document that must accompany a new issue of securities for a real estate company or partnership and must include the same information in the registration statement, such as a list of directors and officers, financial statements certified by a certified public accountant, underwriters, the purpose and use for funds, and other relevant information.

proximity damage. N. Decline in value of real estate property because it is near something that is damaging to its worth.

proxy. N. (1) A person who represents another, particularly in some type of meeting. (2) The document giving a person the authority to represent another.

public auction. N. A meeting in an announced public location to sell property to repay a mortgage that is in default.

public domain. N. Land owned by the federal, state, or county government that the public might use, as distinguished from property owned privately by individuals and businesses. Right to an item belongs to the public at large, so anyone can use it.

public housing. N. Government-owned housing units made available to low-income individuals and families at no cost or for nominal rental rates.

public lands. N. Acreage held by the government for conservation purposes. Public lands are generally undeveloped, with limited activities such as grazing, wildlife management, recreation, timbering, mineral development, water development, and hunting.

public offering. N. Offering of new securities of a real estate company to the investing public, after registration requirements have been filed with the Securities and Exchange Commission.

public record. N. Governmentally held records of public transactions giving constructive notice that documentation exists confirming the transaction. In real estate, deeds, subdivision plats, and assessments that county and municipal clerks store on record cards in a publicly available filing system.

public report. N. A report published by a governmental unit that is publicly available.

public sale. N. A public foreclosure sale where public notice is given and anyone is allowed to participate.

public sewer. See *municipal sewer*.

public syndicate. N. Group of at least two people or businesses combining to engage in a real estate project that would exceed their individual financial abilities. A syndication allows earnings to be proportionately shared.

PUD. ABBRV. Planned unit development.

punch list. N. List detailing items to be fixed, which is compiled by a buyer prior to closing on a property.

punitive damages. N. Used to penalize a defendant for bad faith, malice, fraud, violence, or evil intent, and are designed as both punishment and as a deterrent for future actions of the defendant. Synonymous with *exemplary damages*.

purchase agreement. N. Contract signed by buyer and seller stating the terms and conditions under which a property will be sold, including description of property, price offered, down payment, earnest money deposit, financing, personal items to be included, closing date, occupancy date, length of time the offer is valid, special contingencies, inspections, etc.

purchase money mortgage. N. Mortgage obtained by a borrower as partial payment for a property. Type of seller financing that is a mortgage loan from the seller instead of cash for the purchase price of the real estate. ABBRV. *PMM*. Synonymous with *vendor's lien*.

purchase money transaction. N. The acquisition of property through the payment of money or its equivalent.

purchase order. N. Written authorization to perform a service or supply a material, indicating the cost of such.

purchasing power risk. N. Risk resulting from possible increases or decreases in price levels that can substantially impact real estate values.

PVC piping. N. Piping made from polyvinyl chloride, a lightweight, resilient, chemical-resistant, strong, and durable thermoplastic with a long lifespan, which is often used for cold water systems and in areas where chemicals are found.

pyramid zoning. N. Form of zoning regulation allowing all the uses permitted in more restrictive zoning to also apply to less restrictive zoning. The net effect of pyramid zoning is to pyramid only a few uses to more restrictive zoning regulations while allowing the broader base of uses to be applicable in less restrictive applications.

quadrangle. N. Rectangular plat bordered on all sides by buildings, which is often grassy with decorative landscaping. A quadrangle can be found in a central business district or on the site of an academic institution.

quadrant. N. (1) One of four equal parts created when an object or area is divided by lines that intersect at right angles. (2) One-fourth of a circle.

quadrominium. N. Four-unit building with four tenants in a condominium type of ownership and management.

qualified opinion. N. Accountant's or auditor's opinion of a financial statement for which some limitations exist, such as an inability to gather certain information or a significant upcoming event which may or may not occur.

qualified thrift lender. N. Lender specializing in home mortgage finance under the rules established by the Financial Institutions Reform, Recovery and Enforcement Act of 1989, one requirement being that the lender holds at least 70% of its portfolio in residential mortgage loans and mortgage-backed securities. Qualified lenders are eligible for advances from the district Federal Home Loan Bank.

qualifying. N. Process determining an individual's financial ability to meet the terms of a loan. When selling real estate, the sales broker must qualify the buyer to make certain he or she has the financial ability to purchase the property.

qualifying guidelines. N. There are two main elements lenders consider when determining whether a borrower qualifies for a specific mortgage. The first is the monthly mortgage costs, i.e., principal, interest, tax, and insurance. Mortgage costs should not exceed 28% of a borrower's gross monthly (pre-tax) income. The second qualifying guideline relates to total monthly housing costs and other debts. These costs should not exceed 36% of gross monthly income. Lenders follow these guidelines because they believe these percentages allow homeowners to pay off their mortgages fairly comfortably without the worry of loan defaults and foreclosures. However, these guidelines can be exceeded in certain cases, such as with borrowers who have a good credit history or with a larger down payment.

qualifying ratio. N. Calculations that are used in determining whether a borrower can qualify for a mortgage. They consist of two separate calculations: a housing expense as a percent-of-income ratio and total debt obligations as a percent-of-income ratio.

quantity survey. N. Estimated itemization of all costs in constructing a structure including site acquisition and preparation and a detailed cost estimate of all materials, labor, and overhead required to reproduce a structure. Quantity surveys are used by contractors in preparing a project's bid price.

quantity take-off. N. An itemization of the entire number of items that are necessary to complete a building project as it appears on the blueprint.

quasi-contract. N. Legal obligation to do something imposed upon someone by law, which bears the force of a contract and is subject to legal action as a contract. It is basically a legal obligation to pay for a benefit received as if a contract had actually occurred.

Queen Anne style. N. Victorian-era style of home, which is multi-story and features turrets, high chimneys, and decorative trim.

Queen Anne style

quick assets ratio. N. Balance sheet used to quickly discover the financial health of a business by checking the ratio of current assets without inventories (liquid assets) to current liabilities. A ratio of 1:1 is acceptable. Synonymous with *acid-test ratio*.

quid pro quo. N. (*Latin*) This for that; used to mean something given in exchange for something else.

quiet enjoyment. N. Guarantees the tenant's right to use the premises without interference from the landlord.

quiet title suit. N. Lawsuit filed to ascertain the legal rights of an owner to a parcel of property, to remove a defect, or to remove a cloud on the title.

quitclaim deed. N. Deed that conveys only the grantor's rights or interest in real estate, without stating the nature of the rights and with no warranties of ownership. It is often used to remove a cloud on a title.

quotation. N. (1) The highest bid to buy and the lowest offer to sell a parcel of real estate in a particular market at a specified time. (2) A proposal to perform certain work for a specified price.

R

racial steering. N. The illegal practice of directing certain races away from some neighborhoods and into others.

radiant heating. N. Use of radiation to generate heat such as with baseboard heating, where the circulating hot water is radiated through conduction by thin metal fins at the bottom of the wall. The room is warmed by air circulating around the heating unit using convention.

radon. N. Radioactive gas that seeps into some homes from the ground, through sump pumps, cracks in the foundation, etc.; it is considered a health hazard. See also *contingency*.

rafter. N. Any of the beams that slope from the ridge of a roof to the eaves to serve as support for the roof.

rain cap. See *entrance cap*.

RAM. ABBRV. Reverse annuity mortgage.

rammed-earth construction. N. An alternative building process in which dirt is compacted into large structural frames to create walls.

ranch style. N. Modern style of home that has all the rooms on one floor.

Ranch style

range capacity. N. Total number of range grasslands acres that are needed to support one animal unit for a certain time period.

ratable value. N. An estimated insurance risk to calculate a reasonable premium that would provide sufficient resources, while still being affordable, in the event that the company is required to pay a claim.

rate cap. N. Maximum interest rate charge allowed on the monthly payment of an adjustable rate mortgage during an adjustment period.

rate commitment. N. A written promise by a lender to lend money to a potential borrower at a stated rate of interest for a fixed time period.

rate-improvement mortgage. N. Loan that entitles a borrower to a one-time interest rate cut without refinancing.

rate lock. N. Lender's commitment to borrower to guarantee a specific interest rate for a certain period of time.

rate of return. N. Yield; the return on an investment.

rate sheet. N. A printed card in a hotel that gives the price charged for each room, crib, extra person, and pets.

rate type. N. How payments adjust over a loan term; rate types include fixed-rate, balloon, and adjustable rate.

ratification. N. Approval of a prior act or contract, which gives it the confirmation to make it binding.

rating. N. A value quantifying the capabilities or endurance of an item or substance.

ratio. N. A proportion of one value to another related value. For example, the proportion of defective units to operational units.

rational motives. N. Determination proving that the motivation of a testator was rational when making the devises of a will.

raw land. N. Acreage without added improvement such as sewers, utilities, streets, or structures.

raze. v. To demolish.

ready, willing, and able buyer. N. Real estate term describing one who is capable of action and planning to do so. For example, having the financial ability and being agreeable to the terms of a contract.

real asset. N. An asset that is intrinsically valuable because of its utility, such as real estate or physical equipment.

real capital. N. Capital, such as equipment and machinery, which is used to produce goods. Distinguished from financial capital, which is funds available to acquire real capital.

real estate. N. Land and anything permanently affixed to it, such as buildings. Synonymous with *real property, realty*.

real estate agent. N. Person licensed by a state to represent a buyer or seller in a real estate transaction in exchange for a commission. Agents must work in association with a real estate broker or brokerage company.

real estate attorney. N. A lawyer who specializes in real estate transactions.

real estate board. N. Local group of real estate brokers who are members of the state and national board of realtors. They meet regularly to help determine licensing requirements as well as manage the multiple listing service of their area.

real estate broker. N. Person, corporation, or partnership licensed by a state to represent a buyer or seller in a real estate transaction in exchange for a commission. Brokers supervise licensed real estate agents who act for the broker, who is legally the principal agent in any transaction.

real estate calculators. N. Calculators that have additional financial functions, which include present value, purchase price, property appreciation, lease costs, and loan and mortgage amortization.

real estate commission. N. (1) The amount received by a real estate salesperson and agency upon sale of a property. (2) The agency that enforces real estate license laws.

real estate counselor. N. Person who is paid to provide advice about real estate.

Real Estate Educators Association. N. Professional organization composed of teachers of real estate in colleges and proprietary license preparation schools.

Real Estate Index. N. Data provider for many types of property in more than fifty cities.

real estate investment trust. N. Publicly traded company that owns, develops, or operates commercial properties. ABBRV. *REIT*.

real estate market. N. The current transaction activity by buyers and sellers, including markets for various properties, such as housing, condominium, land, and office markets.

Real Estate Mortgage Investment Conduit. N. Mortgage-backed security that separates mortgage pools into different maturity and risk classes. ABBRV. *REMIC*.

real estate owned. N. Property acquired by a lender through foreclosure, which is held as inventory. ABBRV. *REO*.

real estate property tax. N. Local government taxes that are assessed on real estate.

Real Estate Settlement Procedures Act. N. Federal law designed to make sellers and buyers aware of settlement fees and other transaction-related costs. It also outlaws kickbacks in the real estate business. ABBRV. *RESPA*.

real estate syndicate. N. Pool of investor money, which is used to purchase real estate.

real estate valuation. N. Professional opinion of the market value of a home or property.

real income. N. Income for a certain period, adjusted for inflation.

real interest rate. N. Interest rate adjusted for inflation.

realized gain or loss. N. The difference between the amount received upon the sale or other disposition of property and the adjusted basis of the property.

realized profit or loss. N. Reported on the income statement for tax purposes, it is the taxable profit or loss resulting from a sale.

real property. See *real estate*.

real return on investment. N. Return on an investment after adjustment for inflation.

Realtor®. N. Designation for an agent or broker who is a member of the National Association of Realtors® and subscribes to a strict code of ethics.

Realtor-Associate®. N. Licensed salesperson who is a member of the National Association of Realtors®.

Realtor-National Marketing Institute. N. An affiliate of the National Association of Realtors®, which produces educational programs and literature for its members. Publications include *Real Estate Today* and *Real Estate Perspectives*. ABBRV. *RNMI*.

realty. See *real estate*.

reappraisal lease. N. Lease with a rental, which is a percentage of the appraised value and is periodically adjusted after being reviewed by independent appraisers.

reasonable consent. N. Standard applied in a lease, limiting the landlord's ability to withhold consent in its sole discretion.

reassessment. N. The revision or reappraisal of the value estimate of property, which may be for tax purposes or contract negotiations.

rebate. N. (1) Refund that resulted from an overpayment of tax or purchase price. (2) An amount given as an incentive to buy.

recapture. V. To return an owner's investment through, among other things, depreciation allowance.

recapture clause. N. Clause in a lease that would allow the landlord a percentage of the tenant's profits over the original fixed amount of rent or, alternately, allow the landlord to cancel the lease if the profits of the tenant fall below a specific level.

recapture of depreciation. N. Each year that a depreciable business asset is owned, depreciation is claimed that theoretically corresponds with the using up of the property through normal wear, obsolescence, etc. Thus, the property should be worth approximately its adjusted basis. If the property is sold for more than its adjusted basis, Section 1245 of the Tax Code requires that the gain on personal property and certain nonresidential real property (to the extent of depreciation claimed) be recaptured, i.e., included as ordinary income on the tax return. The purpose of this recapture is to prevent capital gain treatment of gain resulting from claiming depreciation. The recapture of depreciation or cost recovery rules do not apply when the property is disposed of at a loss.

recapture rate. N. Annual return rate of the capital of a wasting asset, which is returned from the depreciating asset's earned income.

recasting. N. Loan term revision that is often made when a borrower is having difficulty making the payments, such as extending the loan for additional years or modifying the interest rate.

receipt. N. The receiving of something or a written acknowledgment that something, such as cash or documents, has been received.

receivables. N. Money owed to a business by customers.

receiver. N. Court-appointed manager of the affairs of a business or piece of property during a bankruptcy or foreclosure. The responsibility for managing the affairs prudently (collecting funds, paying bills, etc.) is carried out under court direction and may either result in a return to a solvent state or a recommendation for liquidation.

recession. N. Business cycle phase of a deteriorating economy, which results in less business and consumer spending and depression of real estate prices. Specifically defined as two consecutive quarters with negative economic growth.

reciprocity. N. Situation in which individuals or entities give certain rights to each other in return for the rights being given to them.

reclaim. V. (1) To convert property from an unusable state (i.e., state that is contaminated, flooded, etc.) to a useful condition. (2) To secure the return of property or rights.

reclamation. N. Conversion of property from an unusable to a usable condition.

recognized gain. N. Taxable income portion of the money received from the sale of real estate.

reconciliation. N. In an appraisal, the process of adjusting comparables for an estimated value of a subject by using cost, market comparison, and income approaches.

reconstructed operating statement. N. Statement of income that is either revised due to new information or reconstructed from information or records if the original is lost.

reconveyance. N. The conveying of a property by a lender back to the borrower once he or she has completely paid off his or her mortgage.

reconveyance deed. N. Deed that is issued to convey a property to the original owner once the mortgage is repaid.

recordation. N. The recording of deeds and other instruments in a public registry to give notice of ownership or legal and financial claims to the public, which protects those making claims from unrecorded claims.

recorded map. N. A plan or map of a specific land area that shows the boundaries of individual properties. Synonymous with *recorded plat*.

recorded plat. See *recorded map*.

recorder. N. Public official who is responsible for keeping records of all real estate transactions.

recording. N. Filing of property-related documents into the public record.

recording fee. N. Fee charged for conveying the sale of a piece of property into the public record.

recourse loan. N. Loan that gives the lender access to additional capital beyond the pledge collateral that secures the loan. If the borrower defaults on the loan, the lender may pursue other assets to recover its loan in full.

recovery fund. N. A fund that is comprised of charges assessed as part of the registration fee for licensed real estate brokers, which is used to compensate individuals who have sustained losses in a transaction with a broker or agent. If a settlement is awarded by the state's real estate commission, the fund's assets will be debited if the broker or agent fails to provide a recovery.

recovery period. N. Period of time, determined by each individual state, in which a person can seek financial recovery from a broker or agent.

recovery property. N. Tangible depreciable property that is not excluded from the accelerated cost recovery system or the modified accelerated cost recovery system. Generally, this property is acquired for use in a trade or business or property held for the production of income.

rectangular land survey. N. A survey used to subdivide public land, which divides a district into twenty-four square mile quadrangles from the meridian (north-south line) and the baseline (east-west line). The tracts are divided into six-mile square parts called townships, which are in turn divided into thirty-six tracts, each 1-mile square parts, called sections. Synonymous with *government rectangular survey*.

redemption period. N. Period of time during which a property owner can pay all defaulted payments and charges and redeem a defaulted mortgage or land contract. The time period varies as established by state statute.

redevelop. V. To rebuild an area with new structures after the demolition and removal of the existing structures.

redlining. N. The overt or covert practice by a bank or insurance company to deny credit or insurance to people based on their ethnic backgrounds or neighborhoods.

reduction certificate. N. Certification in writing by the lender of the remaining balance, date of maturity, and interest rate on a mortgage.

refinancing. N. Modifying existing debts, including mortgages, typically by replacing one or more existing obligations with new loans. Usually done when interest rates are more favorable or when the original debt can no longer be afforded.

reformation. N. Correction of a contractual error that did not reflect the intent of both parties to the deal. Fault needs to extend to both parties unless the error of one person was due to fraud by the other.

regional shopping center. N. Type of shopping center, which is often enclosed and contains three hundred thousand to nine hundred thousand square feet of shopping space, including at least one major department store.

register. V. To formally record a transaction or event.

registrar. N. Individual who maintains official records, such as mortgages, deeds, etc.

registration. N. Notifying a property owner that the property has been shown to a prospect in order to establish a claim to a commission.

registration statement. N. Documented relevant information about a new securities issue of a company or limited partnership, which must be filed with the Securities and Exchange Commission. This lengthy document contains financial, historical, and administrative details about the issue, which allow investors to make educated decisions.

Regulation Z. N. A U.S. Federal Reserve regulation under the Truth in Lending Act, stipulating that a borrower must be advised, in writing, of all the costs and terms connected with any mortgage or other loan.

rehabilitate. V. To restore a building or structure to a good condition.

rehabilitating tax credit. N. Tax Reform Act of 1986 provides incentive for the use and rehabilitation of old structures or historical buildings. This credit, which is based on a percentage of the cost incurred in the rehabilitation, is given in an effort to arrest urban decay.

rehabilitation mortgage. N. Mortgage that provides for the costs of repairing and improving a resale home or building.

reinstatement clause. N. Insurance policy clause, which states that policies that lapse because of nonpayment of premiums can only be reinstated if all unpaid premiums are paid and other requirements are fulfilled.

reinvestment rate. N. Interest rate assumed by investors to be able to be earned on intermediate cash flow in the projection of terminal value.

REIT. ABBRV. Real estate investment trust.

release. N. (1) The freeing of real estate from a lien once the mortgage is paid in full and the debt retired or the forgiveness of the debt by the creditor. (2) The voluntary abandonment of a legal right against another.

release clause. N. Provision in a purchase contract that allows a seller to continue marketing the home and accepting other offers.

release of lien. N. To free a piece of real estate from a mortgage.

relocation. N. The movement of a person or business from one region or location to another.

relocation benefits. N. Benefits provided by employers for new workers, which can include moving costs, reimbursement for temporary housing and transportation, real estate agent assistance, and discounted loans.

relocation clause. N. Clause in a lease allowing the landlord to move a tenant within the same building.

relocation company or service. N. Firm that administers all aspects of relocating new employees from one location to another.

remainder. N. An interest or estate that remains after all costs have been deducted or when the original life tenant has died.

remaining balance. N. The amount of unpaid principal on a home loan.

remaining term. N. Original loan term minus the payments already made.

remediation. N. Cleanup of an environmentally contaminated site.

REMIC. ABBRV. Real Estate Mortgage Investment Conduit.

remodel. V. To update or alter the appearance and functional utility of a building.

renegotiated rate mortgage. N. A fixed-rate mortgage that expires at pre-established times, which allows for renegotiation of the terms of the mortgage. This mortgage comes due in a balloon payment, which may be paid or refinanced at current rates. ABBRV. *RRM*. Synonymous with *rollover mortgage*.

renegotiation. N. Legal revision of the provisions, terms, or conditions of a contract.

renewal. N. Right of a tenant to renew a lease at a rent to be determined.

renewal option. N. A right, without any obligation, of a tenant to continue a lease at a specified term and rent.

renovate. V. To change or upgrade an existing property.

renovation cost. N. Total amount expended to change or upgrade an existing property.

rent. N. Amount paid from a tenant to a landlord for the use of property.

rental agency. N. Business that aids a tenant in finding the best rental property or a landlord in finding a good tenant.

rental concessions. N. Discounts and reductions in rental charges to attract new tenants or keep present ones. Concessions may also be in the form of some free rental or a large allowance to adapt the space to the needs of the tenant. High occupancy will induce a large retailer to relocate or a bank to offer better financing.

rental contract. N. A lease; a contract providing for the payment of rent by the lessee to the lessor, for the use of real property for a stated time period.

rental income. N. Income received by the taxpayer for allowing another person's use of the taxpayer's property. Rental income includes advance rental payments, late payments, and current payments. Payment received for lease cancellation and forfeited security deposits is rental income in the year received or forfeited. Rental income is considered passive income for purposes of the passive loss rules, except for that of qualified real estate professionals.

rental rate. N. Periodic charge for each rental unit for a specified period of time.

rental value. N. Valuation of the worth of a rental property by considering the net income derived from the property and the capitalization rate.

rent bid model. N. Model based on an assumption that space should be controlled by the activity that offers the highest bid. Maximizes usefulness.

rent control. N. Governmental policy that controls the rate that may be charged to tenants for space rented.

renter's insurance. N. Policy for renters that covers their possessions in the case of damages, accidents, or losses.

rent escalator. N. Lease provision that allows the landlord to raise the rental rate to account for inflation or higher interest rates.

rent-free period. N. Portion of the term of a lease when no rent is required, usually as part of a concession.

rent loss insurance. N. Policy covering any loss of rent or rental value in the event that damage renders the property uninhabitable.

rent multiplier. See *gross income multiplier*.

rent roll. N. List of tenants including the lease rent and lease expiration date.

rent-up period. N. Amount of time needed to fully occupy newly constructed properties.

REO. ABBRV. Real estate owned.

reorientation. N. The changing of the market appeal of a property.

repayment plan. N. When a borrower falls behind in mortgage payments, many lenders will negotiate a repayment plan rather than go to court.

replacement cost. N. The amount it would cost to replace an asset at current prices.

replacement cost accounting. N. An accounting method that allows for additional depreciation on some part of the difference between a depreciable asset's original cost and its replacement cost.

replacement reserve fund. N. Money that is set aside from homeowners' assessments to replace common property, such as furniture in a planned development's community room.

replacement value. N. The value of an asset as determined by the estimated cost of replacing it.

repossession. N. The taking back of a house or property by the lender or seller.

request for proposal. N. A document issued by a tenant's agent to an owner that invites him or her to submit a proposal to the tenant for leasing of space. This document outlines the specifics of the lease term, expansion and renewal options, rental rate, tenant improvements, and other allowances to be provided by the owner. ABBRV. *RFP*.

resale value. N. The future value of a piece of property, which can be affected by many factors, including the surrounding neighborhood, school scores, and economic and housing market conditions.

rescission of contract. N. Contract cancellation that is done for certain reasons, such as illegality of the deal.

reserve fund. N. Money set aside by homeowners' associations for major repairs or improvements.

reserve price. See *upset price*.

residential property energy tax credit. N. Prior to 1986, taxpayers were eligible for a credit against the cost of energy-saving devices or renewable energy source property installed in their principal residences. Residential energy credits claimed in prior years must be subtracted from the basis of the residence.

RESPA. ABBRV. Real Estate Settlement Procedures Act.

restructured loan. N. A mortgage in which new terms are negotiated.

retail gravitation. N. The ability of a shopping center to draw business away from other shopping areas. Usually, the larger the center, the greater its power.

retail land developer. N. Individual or corporation that takes a raw piece of land through the approval process with the town and then installs the necessary utilities to turn the property into individual lots ready for construction.

retail lot sales. N. The sale of the developed lots for building homes, office buildings, industrial buildings, shopping centers, etc.

retail property. N. Property to be used by a retail business for the sale of merchandise or services.

retainage. N. Construction contract term for the funds that are earned by the contractor but not paid until some agreed-upon date, such as the completion of the job. This is considered an incentive to complete the job in a timely manner.

retention clause. N. A provision in a contract that allows the client to hold back a portion of payments until the project is complete.

return on assets. N. Earnings divided by total assets. ABBRV. *ROA*.

return on equity. N. Equity yield rate; the rate of return on the equity portion of an investment, taking into account periodic cash flow and the proceeds from resale. Considers the timing and amounts of cash flow after annual debt service, but not income taxes. ABBRV. *ROE*.

return on investment. N. Profit generated by a property, such as rents, etc. Usually stated with the profit as a percentage of the total amount invested. ABBRV. *ROI*.

reuse appraisal. N. Appraisal to determine the resale value of a vacant or improved property in an urban area that is now or will be under development, done in accordance with the National Housing Act.

revaluation. N. The reconsideration of the value or worth of a property.

revaluation clause. N. Clause in a reappraisal lease (a rental in which the payment is a percentage of the appraised value) that is periodically adjusted after being reviewed by independent appraisers.

revenue sharing. N. The splitting of operating profits and losses between the general partner and limited partners in a limited partnership. More generally, the practice of sharing operating profits with a company's employees or sharing the revenues resulting between companies in an alliance.

reverse annuity mortgage. N. Loan available to older owners who have equity in their homes but need cash. A periodic payment is made to the borrower from the lender, thus increasing the loan balance and causing negative amortization. ABBRV. *RAM*.

reverse leverage. N. Negative cash flow; borrowing money at a higher interest rate than the return obtained by investing that money.

reverse mortgage. See *home equity conversion mortgage*.

reversion. N. An interest or estate in which an individual has a fixed interest in the future, such as the remaining part left after obligations are paid, which would revert for distribution; the right of a lessor, upon termination of a lease, to possess leased property.

reversionary factor. N. The mathematical factor used to determine the current worth of future cash flow.

reversionary interest. N. Interest a person has in property that is now held by another. Upon termination of that possession, the property will revert back to the grantor.

reversionary lease. N. Rental agreement, which will only begin after the expiration of the current rental agreement.

reversionary value. N. Estimated value of a property at the expiration of a certain time period.

rezoning. N. Modification of the designation of a parcel or group of parcels on the zoning map, which changes the permitted usage of the area. Changes in zoning are usually requested by individuals or businesses and then approved by the zoning commission of a town. Normally, these changes are only granted if there is no adverse effect on other properties within the area.

RFP. ABBRV. Request for proposal.

rider. N. An amendment or attachment to a contract or a modification to an insurance policy.

ridge board. N. A horizontal board that serves as the apex of the roof structure.

ridge vent. N. A vent located along the ridge board of the roof that allows moisture to escape.

right. N. (1) A just claim to power or privilege; something that belongs to a person by law, nature, or tradition, etc. (2) The claim of a person or entity to property by exercising an option.

right of access. N. The right of a property owner to go to and return from an adjoining street without interference.

right of courtesy. N. The legal right of a spouse to a life estate in all lands owned by a deceased spouse.

right of dower. N. Rights of a widow whose husband died intestate (without a will) to the use of all his lands and possessions for the support of herself and her children. Upon her dissent of the will, the widow is entitled to one-third of all the assets of the estate of her deceased husband. This right is established in most states and significantly altered in others.

right of entry. N. Right to begin usage, for living purposes or construction, of property in the process of being purchased.

right of first offer. N. This right gives a tenant the first option of buying or leasing occupied property if the owner decides to sell or lease.

right of first refusal. N. A right to buy or rent a piece of property, which is given to a person by the owner of the property, before it is placed on the open market.

right of offset. N. Specific clause in lease in which the tenant has the right to deduct from the rent costs that are incurred to the tenant from the landlord.

right of redemption. N. Borrower's right to redeem property taken in foreclosure by immediately paying off the loan balance and any related costs. The right of a bankruptcy debtor to recover personal property under lien by making restitution to the creditor.

right of rescission. N. Right to cancel, within three business days, a contract that uses the home of a person as collateral, except in the case of a first mortgage loan.

right of subrogation. N. Substitution of one entity or person for another if both have the same rights and obligations.

right of survivorship. N. Survival right of a person when that person has an interest in the property of another, such as in a tenancy by the entirety.

right of way. N. When the property owner lets others use or pass through his or her property.

riparian. ADJ. On the bank of a body of water.

riparian owner. N. Owner who has rights to water on his or her land and a reasonable right to water that flows through his or her property from an adjacent site.

riparian rights. N. The right of one owner to use a river, stream, or lake that borders his or her property. Synonymous with *water rights*.

risk. N. Uncertainty or variability; a chance of loss. As to real estate, the fluctuation in sales or profits and the likelihood of declining value.

risk-adjusted return. N. A measure of how much an investment returned in relation to the amount of risk it took on. Often used to compare a high-risk, potentially high-return investment with a low-risk, lower-return investment.

risk-free rate. N. Interest rate on the safest investments, such as federal government obligations.

ROA. ABBRV. Return on assets.

roadbed. N. The base over which a road's paving is installed. The roadbed is usually topped with graded crushed stone.

ROE. ABBRV. Return on equity.

ROI. ABBRV. Return on investment.

ROIC. ABBRV. Return on invested capital.

rollover loan. N. (1) Long-term loan with a guaranteed interest rate for a shorter period with interest renegotiated periodically, at current rates or the extension of a mature loan at current rates. (2) Delay allowed for repayment of principal, by a bank to a debtor with financial difficulties.

rollover mortgage. See *renegotiated rate mortgage*.

roof joist. See *ceiling joist*.

rough-in. N. The installation of plumbing, electrical, and other mechanical systems.

royalty. N. A payment made for the use of property, especially a patent, copyrighted work, franchise, or natural resource. The amount is usually a percentage of revenues obtained through its use.

Rule of 69. N. A rule used to estimate the time it would take to double an investment, which is that a set amount of money invested at a certain percentage rate per period will double in $69 \div$ percentage rate $+ .35$ years.

Rule of 72. N. A rule used to estimate the time it would take to double an investment when earning compound interest by dividing the percentage rate into seventy-two to derive the number of years required to double the principal. For example, an investment that yields an annual return of 20% will double in less than three years.

Rule of 78. N. A method of computing unearned interest used by banks to formulate a loan amortization on installment loans with add-on interest. The number seventy-eight is based on the sum of the digits from one to twelve. This allows a borrower to pay more interest at the beginning of the loan when there is more money owed and less interest as the obligation is reduced. Synonymous with *Sum-of-the-Years'-Digits method*.

run rate. N. The revenues that a company would have in the next twelve months if the current revenue rate remained unchanged. Usually calculated by multiplying the latest quarter's revenues by four.

run with the land. N. Expression that indicates a right or a restriction that will affect all current and future owners.

rural housing service. N. A U.S. Department of Agriculture program that provides financing to farmers and certain borrowers to purchase rural property when other funds are not available.

R-value. N. A construction term that refers to the resistance of insulation to heat loss. The higher the R-value, the slower the rate of heat loss.

S

sale-leaseback. N. A technique in which a seller deeds property to a buyer for a consideration and the buyer simultaneously leases the property back to the seller. This type of transaction is frequently done by retailers for their store properties.

sales-assessment ratio. N. Selling price of a property divided by its appraisal value. If real estate has a selling price of $400,000 but its assessed value is $380,000, the ratio is 1.053.

sales commission. N. The percentage of the selling price that is paid to a real estate broker for his or her services in obtaining a purchaser.

sales comparison approach. N. Method of appraising real estate by making a market comparison of neighboring properties having similar characteristics to ascertain what it could cost to substitute a similar property for the current one.

sales concession. N. Instance where the seller pays a cost that is normally paid by the buyer. Usually done to ensure that the sale will go through.

sales contract. N. Agreement outlining the terms of a purchase, which is signed by the buyer and the seller.

sales deposit receipt. N. Receipt given for a partial payment made on the sale of the property, which is a down payment.

sales expenses. N. Costs incurred during the sale of real estate, such as real estate commission, attorney fees, etc.

sales incentive. N. Extra compensation given to a real estate broker who has surpassed his or her sales quota, which may be a flat fee or a percentage of the extra sales dollars over the quota.

sales kit. N. Literature, samples, and other useful information used by brokers or agents for purposes of demonstration to prospective purchasers.

salesperson. N. Individual employed in selling a product or service.

sales price. N. The amount of money that is paid by a purchaser to a seller for an object that is bought.

sales price list. N. Written list of the prices being asked for homes or office buildings that are for sale.

sales ratio analysis. N. Evaluation of the cause of the difference between the desired selling price of a property and its appraisal value. Reasons could include unexpected deterioration of conditions in the area, quick sale needed, poor appraisal, etc.

sales value. N. Price a property would bring on the open market.

saltbox style. N. Early-American two- or two-and-a-half-story style from the colonial period. The house is square or rectangular with a steep gable roof that extends down to the first floor in the rear of the building.

Saltbox stle

salvage value. N. Estimated value that an asset will have at the end of its useful life.

SAM. ABBRV. Shared appreciation mortgage.

sandwich lease. N. Lease held by a lessee who becomes a lessor by subletting. Typically, the sandwich leaseholder is neither the owner nor the user of the property.

sanitary sewer. N. House drain that carries waste away from the house to a septic system or a municipal sewer system.

satellite cities. N. Subordinate neighborhoods that are tied to an urban area economically.

satisfaction. N. (1) Fulfillment of a person's needs or desires. (2) Discharge of an obligation through payment or rendering of service.

satisfaction of lien. N. Release and discharge of a lien on property after the terms of the lien have been met.

satisfaction of mortgage. N. Written statement by the lender that the buyer of real estate has paid off the entire mortgage.

satisfaction piece. N. Instrument for recording and acknowledging final payment of a mortgage loan.

savings and loan association. N. Financial institution that specializes in originating, servicing, and holding mortgage loans, primarily on owner-occupied, residential property. S&Ls also make home improvement loans and loans to investors for apartments, industrial property, and commercial real estate. Approximately 40% of S&Ls are federally chartered, while the rest are state chartered. Federal charters are members of the Federal Home Loan Bank System (FHLBS). All federally chartered S&Ls are owned by depositors and the word "federal" must appear in their title. State chartered S&L's can be either mutually owned or stock associations. They have optional membership in both the FHLBS and the Federal Savings and Loan Insurance Corporation. ABBRV. *S&L*. See also *thrift institution*.

SBA. ABBRV. Small Business Administration.

scheduled cash flows. N. The mortgage principal and interest payments due to be paid under the terms of the market. This does not include prepayments.

Schedule E. N. Part of Form 1040 that shows income or loss from real estate transactions, including net rental income.

schematic. N. Plan diagram showing a specific design or functional element of a home such as electrical or plumbing.

seasoned loan. N. Loan where the borrower has proven his or her creditworthiness by having consistently made payments.

secondary financing. N. Loan that is subordinate to the primary loan and cannot be satisfied until the primary loan is paid.

secondary mortgage market. N. A market that deals in the buying and selling of existing mortgages.

second deed of trust. N. Deed of trust or mortgage in which the lender subordinates his or her loan to another lender whose priority is first if there is nonpayment by the borrower.

second home. N. A residence that is not one's primary residence. Under current tax law, a taxpayer may deduct interest on two personal residences.

second lien. N. A lien in addition to a first lien and subordinate to that first lien. It can be satisfied only after the first lien has been satisfied.

second mortgage. N. A second loan that uses an already mortgaged house as collateral. The first mortgage has priority over the second loan. Home equity lines of credit represent second mortgages, or second liens, on a property.

Section I of Homeowners' Policy. N. Provides protection for the home, contents, and accompanying structures.

Section II of Homeowner's Policy. N. Provides comprehensive coverage for personal liability and the medical payments and property damage incurred by those other than the insured.

Section 8 Housing. N. Created by the 1974 amendments to Section 8 of the 1937 U.S. Housing Act, it is a federal program in which the government subsidizes much of the rent, in privately owned apartments, on behalf of qualified low-income tenants.

Section 167. N. Section of the Internal Revenue Code that deals with depreciation. Capital improvements made to real property are depreciable.

Section 179. N. An election to treat the cost of certain qualified property as a currently deductible expense rather than as a capital expenditure. A maximum deduction of $20,000 may be claimed for qualified assets placed in service in 2001. Synonymous with *expensing*.

Section 1031. N. This section of the Internal Revenue Code deals with tax-deferred exchanges of certain property. General rules for tax-deferred exchanges of real estate state that the properties must be like-kind property (real estate for real estate), exchanged and held for use in a trade or business or held as an investment.

Section 1034. N. The section of the Internal Revenue Code that applied to the sale of a principal residence before May 7, 1997. It allowed the deferment of gain on a house when a more expensive house was purchased. Section 1034 was replaced by Section 121 for sales after May 6, 1997.

Section 1221. N. Section of the Internal Revenue Code explaining what a capital asset is and is not. Capital assets include real estate

Section 1231. N. Section 1231 assets are depreciable assets and real estate used in a trade or business and held for more than one year. Under certain circumstances, the classification also includes timber, coal, domestic iron ore, livestock (held for draft, breeding, dairy, or sporting purposes), and unharvested crops.

Section 1245. N. Section of the Internal Revenue Code applying to gains from the sale of personal property subject to depreciation. In most cases, the gains are at a capital gains tax rate limited to the amount of accumulated depreciation taken.

Section 1250. N. Section of the Internal Revenue Code that applies to capital gains from selling real estate that has been depreciated for tax purposes. Most buildings must be depreciated using the straight-line method.

secured loan. N. Loan backed by collateral.

secured note. N. Written obligation of a borrower that is backed by collateral in the event of default.

secured party. N. Lender who possesses the collateral of the borrower if the loan is defaulted upon.

Securities and Exchange Commission. N. Federal agency created in 1934 to carry out the provisions of the Securities Exchange Act. Generally, the agency seeks to protect the investing public by preventing misrepresentation, fraud, manipulation, and other abuses in the securities market. ABBRV. *SEC*.

securitization. N. Process of the borrower giving the lender security to obtain the loan.

security. N. The property that will be pledged as collateral for a loan.

security agreement. N. Legal contract in which the lender controls the pledged property being financed. This agreement describes the property and its location. In the event of default, the lender may sell the collateral.

security deposit. N. Prepayment to a landlord to offset any damage that might occur beyond normal wear and tear; considered a damage deposit. Laws in most states require landlords to hold the deposit in a separate account and refund the amount, if no damage is done, within a specified time after termination of the lease.

security instrument. N. An interest in real estate that allows the property to be sold upon a default on the obligation for which the security interest was created. The security instrument is more specifically described as a security deed, a mortgage, or a trust deed.

security interest. N. Interest in real estate in which the real estate serves as collateral.

seed money. N. Funds, often put up by venture capitalists, needed to finance a new business.

see-through building. N. A vacant building. This term originated with the overbuilding of glass office buildings in Houston during the late 1980s. Without tenants or furniture, it was possible to see through such buildings.

seismic load. See *earthquake load*.

self-amortizing mortgage. N. Mortgage that will retire itself through regular principal and interest payments.

self-contained appraisal report. N. A written appraisal report that contains all the information required by the Uniform Standards of Professional Appraisal Practice, with extensive detail.

seller financing. N. Agreement in which the seller provides the financing for a purchase.

seller's market. N. Very strong real estate market in which sellers have the advantage because there are more buyers than properties for sale.

seller take-back. N. An agreement in which the owner of a property provides financing, often in combination with an assumable mortgage.

seller vs. buyer closing costs. N. Buyers and sellers often negotiate who will pay certain closing costs, and the results vary depending on the negotiated deal. In fact, it is not uncommon for a sales agreement to state that either the buyer or seller pays all closing costs. The agreement that a buyer and seller reach must be specified in the sales contract. Negotiations could depend on a variety of factors, including the quality of the home, how long the home has been on the market, whether there are any other interested buyers, and how motivated the seller is to sell the home.

selling agent. N. A broker or salesperson who writes the contract for a buyer in a real estate transaction.

semiannual. See *biannual*.

semi-custom home. N. A home to which changes for some design aspects can be made during construction, but whose basic structure cannot be changed.

senior mortgage. N. Primary mortgage on a property that takes priority over any other liens and is satisfied before any secondary liens.

senior residential appraiser. N. A designation granted by the Appraisal Institute. An SRA may refer to him- or herself as a member of the Appraisal Institute, though not a Member of the Appraisal Institute (MAI). ABBRV. *SRA*.

sensitivity analysis. N. Technique of investment analysis that enables investors to determine variations in the rate of return on an investment property in accordance with changes in a critical factor, such as how much the rate of return will change if expenses rise 5% or rental income drops 10%. It is an experiment with decision alternatives using a what-if approach.

sentimental value. N. Emotional value of property to each particular person.

separable property. N. Property wholly owned by either spouse that was acquired prior to marriage or was received as a gift or inheritance. This property legally belongs to that spouse and cannot be taken away to satisfy a debt against the other spouse or for estate valuation. Synonymous with *separate property*.

separate property. See *separable property*.

service contract. N. Agreement bought by a homeowner for servicing of household items.

servicer. N. An organization that collects principal and interest payments from borrowers and manages borrowers' escrow accounts. The servicer also services mortgages that have been purchased by an investor in the secondary mortgage market.

servicing. N. The collection of mortgage payments from borrowers and related responsibilities of a loan servicer.

servient estate. N. Land subjected to an easement.

setback ordinance. N. Local zoning law or private limitation on how far in feet a structure might be situated from the curb or other appropriate marker.

settlement. N. The final step before a borrower gets the keys to a new home. The meeting is typically attended by the buyer, the seller, their attorneys if they have them, both real estate sales professionals, a representative of the lender, and the closing agent. The purpose of the meeting is to make sure the property is physically and legally ready to be transferred to the buyer. Final closing costs will be paid at this meeting—they generally include a loan origination fee, attorney's fees, taxes, an amount placed in escrow, charges for obtaining title insurance, and a survey. Closing costs vary according to the area of the country. Synonymous with *closing*.

Settlement Cost Booklet. N. A guide published by the Department of Housing and Urban Development that provides an overview of the lending process, which is given to consumers after they complete a loan application.

settlement document. N. Document detailing what has been paid and by whom.

settlement fees. See *closing costs*.

settlement sheet. See *HUD-1 Settlement Statement*.

settlement statement. See *HUD-1 Settlement Statement*.

severalty. See *sole ownership*.

severalty ownership. N. Person's title to real estate giving him or her exclusive power and rights over it.

severance damages. N. Monetary award given by the government to a person whose property has been subjected to condemnation.

sewage tax. N. Levy placed on those who benefit from a sewer system.

sewer. N. A system of pipes, containment, and treatment facilities for the disposal of plumbing wastes.

sewer line easement. N. Easement to build, maintain, and operate a disposal line for sewage.

shakeout. N. The decline in real estate values that occurs during an economic hardship such as a depression or recession. When this occurs, there is normally huge losses for real estate owners, with some declaring bankruptcy.

shared appreciation mortgage. N. Residential loan with a fixed interest rate set below market rates, with the lender entitled to a specified share of appreciation in property value over a specified time interval. Loan payments are set to amortize the loan over a long-term maturity, but repayment is generally required after a much shorter term. The amount of appreciation is established by the sale of the home or by appraisal, if no sale is made. ABBRV. *SAM*.

shared equity mortgage. N. Home loan in which the lender is granted a share of the equity, thereby allowing the lender to participate in the proceeds from resale. After satisfying the unpaid balance of the loan, the borrower splits the remainder of the proceeds with the lender. Shared equity plans often require the lender to buy a portion of the equity by providing a portion of the down payment.

shared equity transaction. N. Situation in which two buyers purchase a property, with one as the resident co-owner and the other as an investor co-owner.

shopping center. N. Collection of retail stores with a common parking area, and usually one or more large department, discount, or food store, and possibly a movie theater. It may or may not be enclosed in a mall.

short form. N. A document, which is condensed into a page or two, that is used in lieu of the longer, more cumbersome document.

short-term capital gain. N. Profit resulting from the sale of an investment that is held for one year or less. Short-term gains are ordinary income and do not qualify for any special tax treatment.

short-term capital loss. N. Loss resulting from the sale of an investment that is held for one year or less. Short-term losses are deducted from current income and do not qualify for any special tax treatment.

silent partner. N. Partner who remains anonymous but has legal rights and obligations.

simple assumption. N. Loan that keeps the original borrower liable in the event that the new borrower defaults on the loan.

simple interest. N. Interest computations that are calculated only on the original principal amount.

simple rate of return. N. Accounting or unadjusted rate of return. The return computed by dividing the anticipated future annual net income by the required investment in real estate.

single-family dwelling. N. House designed for use by one family. Even if other dwellings are attached, as in a town house, they would have separate plumbing, heating, electrical systems, etc. They may be detached housing, town homes, condominiums, or co-operatives.

single-family properties. N. One- to four-unit properties including detached homes, town homes, condominiums, and cooperatives.

single property syndicate. N. Equity investors involved in the purchase of a single property, with the total amount of money raised being used for the purchase.

single purpose agricultural building. N. An agricultural structure used for only one purpose. Examples are milking sheds and greenhouses. Such buildings may be contrasted with multi-purpose structures like barns, which are generally used for a variety of farming purposes.

sinking fund. N. Fund set aside for periodic payments, aimed at reducing a financial obligation taken out to buy real estate, or to accumulate enough funds to buy property or for plant expansion. The principal deposited into the sinking fund earns interest. The total amount accumulated is then used for the desired purpose.

SIOR. ABBRV. Society of Industrial and Office Realtors.

site. N. A plot of land prepared for or underlying a structure or development; the location of a property.

site analysis. N. Evaluation of an area to determine its appropriateness for designated objectives.

site-built home. N. Home constructed on a piece of property chosen by the potential homeowner.

site development costs. N. Total expenses required to make a property suitable for its designated purpose.

site plan. N. Document that describes how a parcel of land is to be improved. It includes the outlines of all structures and site improvements, such as driveways, parking lots, landscaping, and utility connections.

site survey. N. Determination of the measurements of a specific location.

situs. N. Geographic location of land based on its economic significance. The economic attributes of location, including the relationship between the property and surrounding properties, as well as distant points of interest and the linkages to those points. The situs is considered to be the aspect of location that contributes to the market value of a real property.

six-month adjustable rate mortgage. N. This adjustable rate mortgage (ARM) offers a low initial interest rate for the first six months with an interest rate that adjusts every six months thereafter. The rate caps per adjustment can be 1%–2%; the lifetime adjustment caps can be 4%–6%. This type of mortgage may be right for a borrower who anticipates a rapid increase in income over the first few years of the mortgage because it lets the borrower maximize his or her purchasing power immediately. It may also be the right mortgage for someone who plans to live in a home for only a few years. The interest rate is tied to a published financial index. Six-month ARMs are available with terms of ten to thirty years. They can be used to buy one- to four-family, owner-occupied principal residences including second homes, investment properties, condos, co-ops, and planned unit developments. Manufactured homes are also eligible, but must be built on a permanent chassis at a factory and then transported to a permanent site and attached to a foundation. Advantages of a six-month ARM are that it maximizes a borrower's buying power immediately, especially if he or she expects his or her income to rise quickly in the next few years; it lets the borrower select an index that meets his or her financial needs; it is easier to qualify for due to a low interest rate and a 1% or 2% annual rate cap; and, some let a borrower convert to a fixed-rate loan at certain adjustment intervals.

slum area. N. Depressed part of a city or town that might be a high crime area.

Small Business Administration. N. Federal government agency in Washington, D.C., that makes low-interest loans to qualified small businesses. ABBRV. *SBA*.

small claims court. N. Special court for the purpose of providing fast, inexpensive, and informal settlements of small financial claims between a plaintiff and a defendant, in which the parties represent themselves.

Society of Industrial and Office Realtors. N. Organization affiliated with the National Association of Realtors whose members are mainly concerned with the sale of warehouses, factories, and other industrial property. ABBRV. *SIOR*.

Society of Real Estate Appraisers. N. Organization of professionals who value real estate. ABBRV. *SREA*.

soft costs. See *indirect costs*.

soft market. N. Market in which demand has shrunken or supply has grown too quickly, making sales at profitable amounts for a seller more difficult. This type of market is better for purchasers.

soft money. N. (1) In a development or investment, money that is tax deductible. (2) Costs that do not physically go into construction, such as interest, architectural fees, legal fees, etc.

sole ownership. N. Business owned by one person having all the rights and obligations. Synonymous with *severalty*, *sole proprietorship*.

sole proprietorship. See *sole ownership*.

southern colonial. N. Early American architectural style, elaborately built in a symmetrical way, with columns and a colonnade extending across the front of the house. This home typically has three floors and a gabled roof.

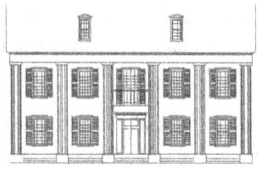

Southern colonial

spec house. See *speculation house*.

special agent. N. Person who is retained to act for another, with limited authority.

special assessment. N. Charge levied against property owners to finance an improvement benefiting the homeowners or commercial businesses. Special assessments may be made by private management boards or local government bodies.

special deposit account. N. An account that is established for rehabilitation mortgages to hold the funds needed as portions of the work are completed.

Specialist in Real Estate Securities. N. Individual who, by his or her expertise, education, and experience, prepares syndication reports. ABBRV. *SRS*.

special purpose property. N. Building with limited uses and marketability, such as a church, theater, school, or public utility.

special use permit. N. Right granted by a local zoning authority to conduct certain activities within a zoning district. Such activities are considered conditional uses, which are permitted within the zone only upon special approval of the zoning authority. Synonymous with *conditional use permit*.

special use property. N. An income-producing property designed for a specific purpose.

special warranty deed. N. Property deed in which the grantor limits the title warranty to the grantee. A grantor does not warrant a title defect to the property occurring from a happening before the time of his or her ownership.

specifications. N. The written requirements for construction, such as for materials, equipment, etc. Synonymous with *specs*.

specific lien. N. Lien on a given property, such as a person's house, as collateral for a loan.

specific performance. N. A legal action in which the court requires a party to a contract to perform the terms of the contract when he or she has refused to fulfill his or her obligations.

specs. See *specifications*.

speculation. N. High-risk, often high-return business transaction undertaken with no guarantee of success.

speculation house. N. A house that is built without a buyer; the dwelling is constructed prior to being sold. The builder "speculates" that a buyer will be found.

speculative space. N. Any tenant space that has not been leased before the start of construction on a new building.

speculative value. N. Value that a speculator believes an investment will reach at some point in the future.

speculator. N. One who invests with the anticipation that an event or series of events will occur to increase the value of the investment.

spendable income. N. Cash flow from income-producing property reduced by income taxes.

split-level home. N. A type of house with floor levels staggered so that each level is about half a story above or below the adjacent one.

Split-level home

spread. N. The difference between the price offered by a buyer and the price asked for by the seller of real estate.

springing power of attorney. N. A type of power of attorney that does not go into effect until a specific event takes place.

sprinkler system. N. A system that dispenses water automatically. There are two types of sprinkler systems; the first is a timer-controlled, in-ground system that waters the grass and shrubbery of a property, and the second is an interior fire protection system that dispenses water from overhead sprinklers in reaction to heat.

square footage. N. Standard unit of area that is used to measure a parcel of real estate; the amount of square feet of livable space in a building or house.

square foot cost. N. Cost of the standard unit of area that is used to measure a parcel of real estate. Commercial property rentals are generally quoted on the basis of square foot cost.

square foot method. N. Means of appraising a building by simply multiplying its square foot cost by the total amount of square feet in the structure being evaluated. Two or more buildings may then be compared by analyzing their total square foot costs.

squatter's rights. N. The legal allowance to use the property of another in absence of an attempt by the owner to force eviction; this right may eventually be converted to title to the property over time by adverse possession, if recognized by state law.

SREA. ABBRV. Society of Real Estate Appraisers.

SRS. ABBRV. Specialist in Real Estate Securities.

stabilization. N. (1) Economic policies designed to reduce the fluctuation in the business cycle, such as those used by the Federal Reserve. (2) The control over rental property exercised by the government in some areas, which places restrictions on amounts able to be charged.

stable mortgage. N. Mortgage loan instrument that combines fixed and adjustable rates in the same loan. The rate applied to the loan is a blend of a fixed rate and a rate that varies according to an index.

stagflation. N. Term coined in the 1970s, increasing prices during a slowdown in economic activity.

standard metropolitan statistical area. N. Designation given by the U.S. Office of Management and Budget to cities of fifty thousand or more residents. ABBRV. *SMSA*.

standard payment calculation. N. The method used to determine the monthly payment required to repay the remaining balance of a mortgage in substantially equal installments over the remaining term of the mortgage at the current interest rate.

standby commitment. N. Commitment by a lender to make available a sum of money at specified terms for a specified period for the financing of a project. The borrower pays a fee for the privilege of either executing the loan or allowing the commitment to lapse.

standby fee. N. The sum required by a lender to provide a standby commitment. The fee is forfeited if the loan is not closed within a specific time.

standby loan. See *bridge loan*.

Starker transaction. See *tax-free exchange*.

starter home. N. A house that is generally of a lower than average price and is often purchased as a first home.

start rate. N. The starting interest rate of an adjustable rate loan. This rate usually lasts between one and twelve months, and then the rate increases based on prearranged criteria. Synonymous with *teaser rate*.

stated value. N. Value assigned to a company's no-par stock by its board of directors, for accounting purposes. Bears no relation to market price.

State Housing Act. N. State law that establishes guidelines and requirements for the construction of buildings. The standard may differ from state to state.

statement of consideration. N. Enumeration of the consideration given by each party to a contract, which in some cases must be in written form to be enforceable. The statute of fraud requires that all contracts for the sale of real estate be in writing.

statute. N. Law established by an act of legislature; can be civil or criminal.

Statute of Frauds. N. A state law that provides that certain contracts must be in writing in order to be enforceable. With respect to real estate, negotiations and preliminary agreements may be oral but the final agreement must be in writing.

Statute of Limitations. N. A specified statutory period in which prosecution or suit must be brought against a person, after which any action will be barred.

statutory dedication. N. Dedication in which the owners of a subdivision or other property file a plat that results in a grant of public property, such as the streets in a development.

statutory foreclosure. N. A foreclosure proceeding not conducted under court supervision.

statutory liens. N. Charges resulting in involuntary encumbrances against real estate derived from legislated law rather than from debts owed to organizations or individuals.

statutory redemption period. N. The limited period of time for a borrower facing foreclosure and sale to attempt to reclaim the property.

statutory right of redemption. N. The legal right of a mortgagor to redeem the property after it has been sold at a foreclosure sale. State law grants the right for a limited period of time, depending on the state.

step-down lease. N. A lease providing for specified rent decreases at certain future dates; opposite of a step-up lease.

step loan. N. Type of adjustable rate mortgage for which the interest rate is adjusted only once during the term of the loan. Therefore, the loan shares some of the features of both fixed rate and adjustable loans.

stepped-up basis. N. An increase in the income tax basis of a property that is a result of a tax-free exchange. As a result of an inheritance, for example, the basis of the inherited property is stepped up to its current market value.

step-rate mortgage. N. A mortgage that allows for the interest rate to increase according to a specified schedule—seven years, for example—resulting in increased payments. At the end of the specified period, the rate and payments will remain constant for the remainder of the loan.

step-up lease. N. Lease that incorporates increases in agreed-on payments over the term of the lease contract. For example, a particular step-up lease may require that the lessee pay a 10% increase each year over the five-year term of the lease. See also *step-down lease*.

stigmatized property. N. Property with an undesirable reputation, which can be due to problems on the site or nearby.

stock of housing. N. Number of housing units of a particular category that are available.

stop clause. N. Provision in a lease agreement that indicates the maximum amount of operating expenditures that must be incurred by the landlord in a given year. Any amount incurred in excess of this amount must be paid by the lessee.

stop notice. N. Notification to a bank that it should not pay a check when presented at the bank; there is a charge for this.

stop work order. N. Written statement from a building inspector to halt work on a project until stated corrections are made.

straight lease. See *flat lease*.

straight-line depreciation. N. Method of depreciation that assumes an asset would lose an equal amount of value each year during its useful life.

straight-line recapture rate. N. Capitalization rate used to convert the expected income derived from a property into its estimated asset value, i.e., capitalized value. The estimated asset value may be computed by dividing the annual income generated by a property by its capitalization rate. The capitalization rate that is used is generally viewed as having two components: a rate of return on investment, and a straight-line recapture rate that represents the percentage of cost that the investor believes he or she must recover each year in order to recoup the entire cost of the asset over its useful life.

straight mortgage or deed of trust. N. Mortgage in which ownership of real estate is placed with one or more trustees as security until a loan is repaid by the debtor and is used instead of conventional mortgages in some states. The deed of trust stipulates that, in the event of default, the trustee would liquidate the property for the benefit of the lender in a trustee's sale.

straight note. N. Loan agreement requiring only interest payments over the term of the loan with a balloon payment of the entire debt balance due at the end of the term.

straw man. N. Individual who purchases property for another individual for the purpose of concealing the identity of the true acquirer from the seller and any other interested parties.

street improvement. N. Repairs made to a street for purposes of safety and attractiveness. In some localities, the homeowner is responsible for properly maintaining the street surrounding his or her home.

strict foreclosure. N. A foreclosure proceeding in which the mortgagee has the right to possess the mortgaged property directly upon default on the mortgage agreement. This type of foreclosure is rarely used in contemporary markets.

strip development. N. Form of commercial land use in which each establishment is afforded direct access to a major thoroughfare; generally associated with intensive use of signs to attract passers-by. Generally, there is no anchor tenant.

structural analysis. N. Evaluation of the dimensions of a building to determine its ability to meet the needs of the occupant.

subagent. N. An agent who assists another agent in representing a principal in a transaction.

subchapter S corporation. N. Corporation with a limited number of stockholders (thirty-five or less) that elects not to be taxed as a regular corporation and meets certain other requirements. Income is taxed as direct income to the shareholders. Shareholders include in their personal tax returns their pro-rata share of capital gains, ordinary income, tax preference items, etc.

subcontract bids. N. Bids that are given in an effort to obtain a job doing work for the general contractor or owner by specialists in certain fields, such as plumbing, roofing, electrical, etc.

subcontracting. N. Work done for the general contractor or owner by specialists in certain fields, such as plumbing, roofing, electrical, etc.

subcontractor. N. Specialists in the construction business who are hired by the general contractor. These would include roofers, plumbers, electricians, etc.

subdivider. N. One who partitions a tract of land for the purpose of selling the individual plots. If the land is improved in any way, the subdivider becomes a developer.

subdividing. N. Dividing a tract of land into smaller tracts.

subdivision. N. A large parcel of property that is divided into smaller pieces.

subject property. N. Appraisal term for the property that is being appraised.

subject to. V. Applies to property that is purchased with conditions to be met, such as "subject to being allowed to be subdivided into a certain amount of lots".

subject to mortgage. N. Circumstance in which a buyer takes title to mortgaged real property but is not personally liable for the payment of the amount due. The buyer must make payments in order to keep the property; however, with default, only the buyer's equity in that property is lost. The buyer is not liable for the amount due to the lender.

sublease. N. Lease agreement between the lessee of an original lease and a new lessee. The new lessee is the subtenant because he or she is renting from the original tenant rather than the owner.

sublessor. N. The initial lessee of rented property who then leases it to a subtenant.

subletting. N. Process by which a lessee leases his or her property to another lessee.

submarket. N. Accumulation of housing units deemed substitutable by homogeneous households, such as those having comparable attractiveness and usefulness.

subordinate. V. To reduce the priority of payment of a debt or lien.

subordinate financing. N. Any mortgage or other lien that has a priority that is lower than that of the first mortgage.

subordinate ground lease. N. A lease where the mortgage has priority over the ground lease.

subordinate loan. N. A second or third mortgage on a piece of property, which already has a first mortgage.

subordinate mortgage. N. A mortgage that has a lower priority than another. The subordinate mortgage has a claim in foreclosure only after satisfaction of mortgages with more priority.

subordination. N. Moving to a lower priority, as a lien would if it changed from a first mortgage to a second mortgage.

subordination clause. N. Clause or document that permits a mortgage recorded at a later date to take priority over an existing mortgage.

subpoena. N. Writ issued by the court requiring a person to appear as a witness or to provide written information in a case. Failure to observe the subpoena may result in a contempt of court citation.

subpoena duces tecum. N. Legal order for a person to present at a deposition or trial documents in his or her possession, such as those related to a real estate transaction.

subprime loans. N. Loans that are made to businesses and individuals who do not qualify for prime rates, i.e., rates reserved for borrowers with virtually blemish-free credit histories. Subprime loans may be anywhere from two to eight points over the prime lending rate.

subrogate. N. The substitution of one person for another with the substituted person acquiring all rights.

subrogation rights. N. Rights allowing an insurer to act against a negligent third party (including its insurance company) to receive reimbursement for payments made to an insured.

subsequent. N. Something that occurs at a later date; it follows a prior occurrence.

subsequent rate adjustments. N. The period for rate adjustment on an adjustable rate mortage after the initial adjustment period. It could differ, in time, from the original duration period.

subsequent rate cap. N. Specific limit for the maximum amount the interest rate may increase at each regularly scheduled rate adjustment date, which could differ from the original rate cap.

subsidiary. N. Being in a secondary or subordinate relationship. ADJ. Of lesser importance.

subsidized housing. N. Housing that has reduced rental payments because of aid granted by the government, private enterprises, or individuals.

subsidy. N. A grant of money made by the government to a private enterprise or another government.

substitution. N. In valuing real estate, the principle that the market value of a property can be relatively accurately estimated by determining market value of similar properties in the general vicinity. By substitution, an appraiser can ascertain the market value of a piece of real estate by analyzing the sales prices of comparable units in the neighborhood sold in the recent past.

subsurface easement. N. Easement in which an owner of land allows another to use space under the ground, such as to install a sewer or gas line.

subsurface exploration. N. Engineering tests performed on soils to determine conditions and ground stability prior to building.

subsurface rights. N. Rights to the soil and minerals underneath the land, which the owner of the real estate generally has exclusively. Significant limitations on subsurface rights have to be disclosed in the title deed at the time of acquisition. If no restrictions are indicated, the buyer can expect to exercise full rights to the property. In many areas, subsurface rights can be extremely valuable because of the existence of oil, natural gas, and minerals, which must be acquired separately.

succession. N. Transfer of real estate by legal means such as through inheritance.

successor. N. Individual coming later in a sequence.

suitability standards. N. Financial characteristics or standards that a potential investor is evaluated on to judge his or her suitability for a particular investment program.

Sum-of-the-Years'-Digits method. ABBRV. *SYD*. See *Rule of 78*.

summary appraisal report. N. A written appraisal report that contains a moderate amount of detail.

summary possession. See *eviction*.

summary proceeding. N. Way to obtain a faster decision in a legal case than going to a trial. Procedural rules are followed so there is less time involved in gathering the facts of the dispute and in questioning.

summons. N. Notice sent from a plaintiff to a defendant requiring the defendant to appear before a court or judge.

sump. N. A drainage system in one's basement that collects excess moisture.

superadequacy. N. A component of real estate that is beyond what is needed in the structure.

Superfund. N. The commonly used name for the Comprehensive Environmental Response, Compensation, and Liability Act (CERCLA), the federal environmental cleanup law. A site on the Superfund list must be cleaned up by any and all previous owners, operators, transporters, and disposers of waste to the site. The federal government will clean such sites as long as the responsible parties pay for the job. Owners who fail to pay such costs risk prosecution.

super regional center. N. Shopping center larger than a typical regional mall.

surety. N. One who guarantees the performance of another, such as agreeing to pay the debts of another if that person does not.

surface rights. N. The right to use and modify the surface area of real estate. In Texas, surface rights are subservient to mineral rights. In the absence of an agreement to the contrary, an oil driller who owns the mineral rights may operate anywhere on the surface of the property.

survey. N. A drawing or map showing the precise legal boundaries of a property, the location of improvements, easements, rights of way, encroachments, and other physical features. A lender may require a borrower to have a survey of the property performed. This process confirms that the property's boundaries are correctly described in the purchase and sale agreement. The survey may show violations, such as a neighbor's fence located on the seller's property. These violations must be addressed before the lender will proceed. The buyer usually pays to have the survey done, but some cost savings may be found by requesting an update from the company that previously surveyed the property.

surveyor. N. One who is trained in the measurement of land for the purpose of determining its perimeter boundaries, contours, and area.

survivorship. N. The right of a joint tenant or tenants to maintain ownership rights following the death of another joint tenant. Survivorship prevents heirs of the deceased from making claims against the property.

sweat equity. N. Value added to a piece of property, by virtue of the work done by the owner, such as in a do-it-yourself improvement.

sweat equity loan. N. Loan given on the premise that the purchaser personally does some work on the property.

sweetener. N. Something included in a transaction to make it more acceptable.

swing loan. See *bridge loan.*

Swiss chalet. N. A one-and-a-half to two-story gable roof house with decorative woodwork in the Swiss style.

Swiss chalet

SYD. ABBRV. Sum-of-Years'-Digits.

symbol schedule. N. Legend that defines and gives meanings for the symbols on a construction drawing.

syndicate. N. (1) A group of investment bankers underwriting and distributing a new or outstanding issue of securities of a real estate business. (2) A professionally managed limited partnership investing in different types of real estate.

syndication. N. A method of selling property whereby a sponsor or syndicator sells interests to investors. It may take the form of a partnership, limited partnership, tenancy in common, corporation, or subchapter S corporation.

syndicator. N. Sponsor of a syndicate involving people or companies buying an interest in a real estate investment or unit. The group of investors are in effect engaged in a joint venture for profit.

synthetic lease. N. A transaction that appears as a lease from an accounting standpoint, but as a loan from a tax standpoint.

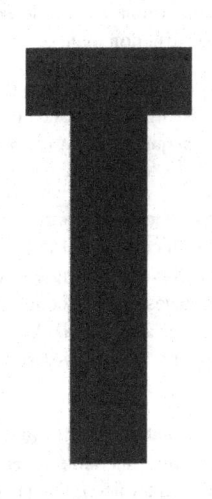

table funding. N. Originating mortgage loans with internal capital until a group of loans can be packaged for sale in the secondary market.

takeout loan or financing. N. A commitment to provide permanent financing following construction of a planned project. The takeout commitment is generally predicated upon specific conditions, such as a certain percentage of unit sales or leases, for the permanent loan to takeout the construction loan.

taking. N. Act of obtaining land through condemnation. Usually the government will exercise its right to take only after it is determined that the owners of the acquired property are unwilling to settle for a reasonable price.

tandem plan. N. Government program of providing low-interest-rate mortgages to low-income qualified buyers. In the tandem program, the Federal National Mortgage Association purchases low-interest-rate mortgages at a discount from the Government National Mortgage Association (GNMA). In doing this, GNMA subsidizes the low-income homebuyer and incurs a loss on the transaction.

tangible asset. N. Asset that has a physical existence, such as cash, equipment, and real estate. Accounts receivable are also usually considered tangible assets for accounting purposes.

tangible property. N. Items of real estate and personal property that usually have a long life, such as housing and other real estate.

tap fees. N. Fees charged for hooking up utilities.

tax abatement. N. A reduction or an exemption from taxes granted by a local government on a piece of real estate for a certain length of time.

taxable income. N. (1) For the individual, adjusted gross income less itemized deductions and personal exemptions. After taxable income is computed, the tax to be paid can be determined by looking at the tax rate schedules. (2) For corporations, taxable income is gross income less allowable business deductions.

tax and insurance escrow. N. A fund established by a lender when a mortgage loan is provided to a borrower to accumulate the debtor's monthly payments for property taxes and insurance premiums for the mortgaged property.

tax assessor. N. Government official who values real estate property for tax purposes and ascertains the annual property tax assessments that must be collected.

tax avoidance. N. Payment of the minimum tax by using legal tax planning opportunities, such as estate planning.

tax base. N. The collective value of property, income, or other taxable activity or assets subject to a tax. Tax revenues are computed as the tax base times the tax rate. For property taxation, the tax base is the total assessed value of all taxable property less exemptions.

tax basis. N. The cost basis of property, such as a home owned for tax purposes.

tax benefit rule. N. A rule that limits the recognition of income from the recovery of an expense or loss properly deducted in a prior tax year to the amount of the deduction that reduced taxable income.

tax book. N. Compilation of all tax maps of a given tax district that are bound together and kept at the local tax office. The tax book is a public record that an individual can review upon request.

tax credit. N. A reduction against income tax payments that would otherwise be due.

tax deduction 358

tax deduction. N. Tax break that is allowed by the government and used to reduce taxable income. Items include mortgage interest, loan points that are paid, and property taxes.

tax deed. N. The type of instrument given to a grantee by a government that had claimed the property for unpaid taxes.

tax district. N. Region or locality that assesses real estate taxes on the properties located within its borders. Frequently, the local county or city is the property-taxing jurisdiction that is empowered, by law, to assess taxes on real estate.

taxes and insurance. N. The tax and insurance components of a mortgage payment are generally held by the lender in an escrow account. The lender pays any property tax and homeowners' insurance bills as they are due, ensuring they are paid on time. These bills may be paid separately by the owner or paid through an escrow fund set up by the bank or lending agency.

tax evasion. N. Failure to pay taxes legally due the government, often by failure to report some income received or claiming false deductions.

tax-exempt. N. Tax-free status given to certain nonprofit organizations and governmental entities.

tax-exempt bond. N. Bond with an interest that is free of federal, state, or local tax in the state of the issuer, and is typically a municipal bond of a state or county agency.

tax-exempt property. N. Real estate that is not subject, in whole or part, to ad valorem property taxes. Churches, charities, and government buildings do not pay property tax because of their tax-free status.

tax flaw. N. Defect in the tax law that either may provide a loophole to minimize the tax payment or result in higher taxes than there should be.

tax foreclosure. N. Property taken over by the government because the owner has failed to pay taxes on it. The property may revert back to the owner when the taxes are paid; if not, the government may sell the property to collect the amount due.

tax-free exchange. N. Transfers of property specifically exempt from federal income tax consequences in the current year; for example, a transfer of property to a controlled corporation. Synonymous with *like-kind exchange, 1031 exchange, Starker transaction.*

tax home. N. The business location, post, or station of the taxpayer. If an employee is temporarily reassigned to a new post for a period of one year or less, the taxpayer's tax home is his or her personal residence and the travel expenses are deductible.

tax impound. N. Money paid to and held by a lender for annual tax payments.

tax liability. N. The amount of total tax due to the Internal Revenue Service after any credits and before taking into account any advance payments—withholding, estimated payments, etc.—made by the taxpayer.

tax lien. N. Legal claim placed on a property for nonpayment of taxes.

tax map. N. Map that documents the area, perimeter location, dimensions, and other data relating to land for purposes of assessing annual real estate taxes.

tax participation. N. Standard portion of a lease, which indicates that, in the event of an increase in annual assessed real estate taxes during the lease term, the lessee will be responsible for higher monthly payments to cover the increase.

tax planning. N. Evaluation of different tax options of investing in real estate with the purpose of reducing the tax liability in current and future years.

tax preference item. N. Item falling under Section 57 of the Internal Revenue Code that may result in the imposition of the alternative minimum tax. These items of otherwise exempt income or deductions, or of special tax benefit, were targeted to ensure that taxpayers who benefit should pay the minimum amount of tax.

tax rate. N. The ratio of a tax assessment to the amount being taxed; amount of tax to be paid based on taxable income. The tax rate usually changes as the unit of the tax base changes.

tax refund. N. Amount the taxpayer gets back when he or she files the tax return at the end of the reporting year because taxes were overpaid for that year. The tax overpayment equals the tax payments remitted less the tax actually due.

tax return. N. Generic name of the form used to file taxes payable to a federal, state, or local government. The tax return includes items such as gross income, deductions, tax credits, and tax due. Individual taxpayers file on a calendar year basis using Form 1040, which is due three and a half months after the tax year, while corporations prepare Form 1120 on a calendar year or fiscal year basis, which is due two and a half months after the tax year.

tax roll. N. The list of all properties subject to a tax in a county or other property-taxing jurisdiction. It identifies all properties and indicates their assessed values.

tax sale. N. Public sale of property by the government for nonpayment of taxes.

tax shelter. N. Term that refers to various tax-advantaged investments and transactions—real estate is considered a good way to shelter income from taxes.

tax shield. N. Deductions that result in a reduction of tax payments. The tax shield equals the amount of the deduction times the tax rate.

tax stop. N. Clause in a lease that sets a maximum limit on the amount of property taxes that a lessor will pay. The lessee is required to pay any taxes in excess of that limit.

teardown. N. Building in such poor condition that it cannot be repaired and must be completely rebuilt.

teaser rate. See *start rate*.

tenancy. N. The right of possession of real estate, whether by ownership or rental; the period for which a tenant has the right of possession.

tenancy at sufferance. N. Tenancy established when a person who had been a lawful tenant wrongfully remains in possession of property after expiration of a lease.

tenancy at will. N. A license to use or occupy lands and buildings at the will of the owner. The tenant may decide to leave the property at any time or must leave at the landlord's will. Agreement may be written or oral.

tenancy by the entirety. N. A tenancy in which parties jointly own property. After the death of one party, the survivor takes the whole estate. Tenancy by the entirety can be terminated during their lifetimes only by joint action of the parties.

tenancy for life. N. A freehold equity in an estate, restricted to the duration of the life of the grantee or other stipulated individual.

tenancy for years. N. A tenancy that lasts for a fixed term.

tenancy from year to year. N. Possession and use of a property estate by virtue of a lease. There are four types of leasehold estates: estate for years, periodic tenancy, tenancy at will, and tenant at sufferance.

tenancy in common. N. A tenancy in which two or more individuals jointly own property. Each owns an undivided share of the whole, and the shares remain separate, even if one party dies.

tenancy in severalty. N. Ownership of property by one person or one legal entity (e.g., corporate ownership).

tenant. N. Individual or corporation renting a residential or office unit. See also *lessee*.

tenant changes. N. Changes made by a lessee to property during the term of the lease.

tenant fixtures. N. Fixtures added to leased real estate by a lessee that, by contract or by law, may be removed by the lessee upon expiration of the lease.

tenant improvements. N. Changes, typically to office, retail, or industrial property, to accommodate specific needs of a tenant. Includes moving interior walls or partitions, carpeting or other floor covering, shelves, windows, toilets, etc. The cost is negotiated in the lease.

tenant-stockholder. N. A tenant in a cooperative share loan, who is both a stockholder in a cooperative corporation and a tenant of the unit under a proprietary lease or occupancy agreement.

tenants' union. N. Group of rental occupants acting together.

tender. V. To present something of value for another's acceptance.

tenement. N. A city apartment building that is overcrowded, poorly constructed or maintained, and generally part of a slum.

tenure. N. The length of time or condition under which something, such as a piece of property, is held. The nature of an occupant's ownership rights; an indication of whether one is an owner or a tenant. See also *land tenure*.

term. N. (1) The period of time during which something is in effect. A stated number of years. (2) A condition or arrangement specified in an agreement. See also *amortization term*.

term beginning. N. The statement on a lease citing the date that the tenant must start paying rent.

termination clause. N. Provision in a contract that upon a certain occurrence or event the contract is cancelled.

termite clause. N. A provision in a sales contract that allows the buyer to have the property inspected for termite infestation. In general, if termites are discovered, the buyer may require the seller to treat the property or the buyer may cancel the contract. Most clauses now use wood-destroying insects to protect against other types of insects that harm structures.

termite inspection. N. Examination of a property for both visible termite infestation and termite damage. See also *contingency*.

termite shield. N. Aluminum- or metal-treated barrier that is placed between the concrete and wood of the foundation of a newly constructed building to prevent termites from infesting the wood. Many experts say that such a shield is ineffective because the termites simply go around the shield to get to the wood in the foundation.

term length. N. The effective period of a lease.

term loan. N. Loan requiring only interest payments until the last day of its term, at which time the full payment is due.

terrace. N. An unroofed paved area right next to a house, a roofed balcony, a veranda, or a raised bed of earth constructed to enhance a landscape.

testament. See *will*.

testamentary capacity. N. The ability to comprehend the terms and conditions of a will and their impact. A testator must understand his or her estate and its eventual disposition and effects in order to validate the will.

testamentary intention. N. Interpreting the objective of a testator in his or her will. The disposition of the testator's estate must be understandable or it could be legally challenged.

testamentary trust. N. Trust created by a will, which comes into effect only after the death of the testator. It empowers a trust administrator to implement the terms of the trust.

testate. ADJ. Having made a legally valid will.

testator. N. Person who makes a legally valid will.

testimonium. N. A clause that cites the act and date in a deed or other conveyance. Before signing a deed, the grantor should make sure that everything is in order.

thin market. N. Market in which there are comparatively few bids to buy or offers to sell real estate. The term relates to a single investment or a particular investment market, such as the real estate market. In a thin market, buying or selling a few homes can make a disproportionate impact. Price volatility is generally wider than liquid markets.

third party. N. One who is not directly involved in a transaction or contract but may be involved or affected by it.

third-party origination. N. A process by which a lender uses another party to completely or partially originate, process, underwrite, close, fund, or package the mortgages it plans to deliver to the secondary mortgage market.

threat of condemnation. N. When there is reason to believe that a property will be taken by the government, usually due to a property owner being informed, either orally or in writing, by a representative of a governmental body or a public official authorized to acquire property for public use, that a decision has been made to acquire his or her property.

three-dimensional shingles. See *laminated shingles*.

three-day notice to quit. See *notice to pay rent or quit*.

thrift institution. See *savings and loan association*.

tight market. N. Market in which the spreads between the asking and offering prices of real estate are small. The property may be in abundant supply and actively traded.

TILA. ABBRV. Truth in Lending Act.

time and materials contract. N. A contract providing for the contractor to be paid for time spent on the job plus the cost of materials.

timeshare. N. A kind of property ownership that allows an owner to occupy a property during a specific period of time, or that percentage of interest, in a vacation home or resort.

time value of money. N. A concept that money available now is worth more than the same amount in the future because of its potential earning capacity. It is the rationale behind compounding for future value or discounting for present value.

title. N. Legal document that shows ownership of a piece of real estate.

title block. N. Markings on a blueprint containing design and revision information.

title charges. N. Fees related to the transfer of title, which typically include closing fees, title insurance premiums, title searches, document preparation fees, and attorney fees. The fees the buyer pays for a real estate attorney are not part of settlement procedures. See also *buyer costs*.

title company. N. Firm that provides insurance of a clear title once it completes its search for liens.

title defect. N. An unresolved claim against the ownership of property, which prevents presentation of a marketable title. Such claims may arise from failure of the owner's spouse or former partner to sign a deed, current liens against the property, or an interruption in the title records to a property.

title examination. N. The examination of the public record by the title company to determine the legal ownership and make sure that there are no claims or liens affecting the property.

title insurance. N. Insurance that protects the lender and the buyer against loss arising from disputes over ownership of property. A lender will require the purchase of title insurance to ensure the borrower is receiving a clear, marketable title. There are two types of title insurance policies: the mandatory lender's policy, which protects the lender should a flaw in the title be detected after the property has been purchased, and the optional owner's policy, which protects the borrower exclusively if a flaw in the title is detected after the property has been purchased. Generally, the buyer pays the cost of both policies.

title insurance binder. N. Written commitment issued by the title company agreeing to insure title to a property, subject to conditions or exclusions shown.

title report. N. A document indicating the current state of the title, including easements, covenants, liens, and any defects. The title report does not describe the chain of title.

title risk. N. Possible impediments to the title transfer of a property.

title search. N. A review by a licensed title company of the title records on a particular property to ensure that the seller is the legal owner of the property and that there are no liens or other claims outstanding before the property passes to a new owner. It attempts to uncover any encumbrances on the title and makes sure the seller is the actual owner of the property. Encumbrances include any liens—legal claims against a property filed by creditors as a means to collect unpaid bills. Liens can also be filed by the Internal Revenue Service for nonpayment of taxes. Any such claims must be paid by the seller—this often occurs either before or at the closing.

title sheet. N. The first page of a set of construction drawings. Information on the title sheet typically includes the name and address of the architect, as well as an index to the plans.

title theory state. N. State in which the law splits the title of mortgaged property; legal title is held by the lender and equitable title is held by the borrower. This is based on the legal assumption that the mortgagee is a partial owner of the real estate securing the mortgage and remains as is until the debt is fully paid; he or she gains full title to the property upon retiring the mortgage debt. The lender is granted more immediate cure to a default than in lien theory states.

to have and to hold clause. See *habendum clause*.

topographic map. N. A map with contours showing changes in elevation.

topographic symbols. N. Symbols and markings used to represent terrain features on a topographic map.

topography. N. Art of mapping the physical features of a region. The topography describes the characteristics of an area, such as its contours and flatness.

top producer. N. Term used in the real estate community to identify agents and brokers who sell a high volume of properties.

topsoil. N. The top layer of soil that is removed when lots are graded in preparation for construction.

Torrens System. N. A title registration system used in some states. Named after Sir Robert Torrens, a British administrator of Australia, this system allows the condition of a title to be discovered without resorting to a title search.

tort. N. A wrongful act that is neither a crime nor a breach of contract, but that renders the perpetrator liable to the victim for damages.

tortfeasor. N. One who has committed a tort.

tort liability. N. Legal obligation stemming from a civil wrong or injury for which a court remedy is justified. A tort liability arises because of a combination of a direct violation of a person's rights, the transgression of a public obligation causing damage, or a private wrongdoing.

total expense ratio. N. The percentage of monthly debt obligations relative to gross monthly income.

total loan amount. N. Amount of money borrowed plus any financed closing costs.

total monthly housing costs. N. Total costs made up of principal, interest, property taxes, and insurances.

total paid at closing. N. All closing costs, which includes the down payment and any prepaid fees. Additionally, two months worth of housing expenses may be required.

total return. N. Return earned on an investment over a given time period. It includes two basic components: the current yield, such as rental income and capital gains, and losses in property values. It is typically stated as an annual percentage.

townhouse. N. A dwelling unit, generally having two or more floors plus a garage, which is attached to other similar units via shared walls. Such dwellings are typically found in condominiums and cooperatives, or as part of a planned unit development. In a townhome development, owners are likely to share assessments for common area maintenance and various amenities.

Townhouse

township. N. Six-mile by six-mile square area of land delineated by a government rectangular survey.

township lines. N. Lines determined by a government rectangular survey laying out a standard six-mile square area of land.

tract. N. A parcel of land, generally held for subdividing and development into residential units.

tract home. N. A mass-produced house constructed within a project by the same builder, with a similar style and floor plan as the other homes in the development.

tract index. N. System for listing recorded documents affecting a particular tract of land.

trade equity. N. Other assets, which could include real estate, that are given by a buyer to a seller as part of the down payment.

trade fixture. N. Articles of personal property installed in rented buildings by the tenant to help carry out a trade or business. Trade fixtures are removable by the tenant before the lease expires and are not true fixtures.

trading down. N. Buying a less expensive home than the one owned currently.

trading on equity. See *financial leverage*.

trading up. N. Buying a more expensive home than the one owned currently.

traffic circle. N. A circular intersection allowing for continuous movement of traffic at the meeting of major roadways.

transaction. N. (1) An agreement between a buyer and a seller to exchange an asset for payment. (2) In accounting, any event or condition recorded in the books of account.

transaction broker. N. A real estate professional who does not represent either the buyer or the seller, but is hired to help them reach an agreement.

transaction costs. N. The costs associated with buying and selling real estate.

transfer. v. To convey something from one entity or person to another.

transfer development rights. n. A type of zoning ordinance that allows owners of property zoned for low-density development or conservation use to sell development rights to other property owners. The development rights purchased permit the landowners to develop their parcels at higher densities than otherwise. The system is designed to provide for low-density uses, such as historic preservation, without unduly penalizing some landowners.

transfer fee. n. Fee charged by a mortgage lender to a buyer, seller, or both for transferring a mortgage when the mortgage property is sold.

transfer of ownership. n. Any means by which the ownership of a property changes hands. Lenders consider the purchase of a property subject to the mortgage, the assumption of the mortgage debt by the property purchaser, and any exchange of possession of the property under a land sales contract or any other land trust device as transfers of ownership.

transfer tax. n. State or local tax payable when title passes from one owner to another.

transom. n. A small hinged window directly above a door.

TransUnion. n. One of the "Big Three" credit-reporting bureaus that operate nationwide. The other two are Experian and Equifax.

tray ceiling. n. A ceiling with edges that slant toward the middle from the walls.

Treasury bills. n. Securities issued by the Treasury Department that have the full backing of the U.S. government.

Treasury Index. n. Index used to determine interest rate changes for adjustable rate mortgages.

trellis. N. A decorative landscape structure made of thin strips of wood or plastic.

trilevel. N. Popular style of home, best suited for side-to-side slopes, in which a one-story wing is attached between the levels of a two-story wing.

trim work. N. The finishing of doors, doorways, window frames, and floors.

triple net lease. ABBRV. NNN. See *net lease*.

triplex. N. A freestanding building having three separate housing units all under one roof.

trophy building. N. A landmark property that is well-known by the public and highly sought by institutional investors, pension funds, and insurance companies. Generally one-of-a-kind architectural designs, with the highest quality of materials and finish, and expensive trim. These properties are more desirable than class A buildings.

truss. N. A prefabricated framework of girders, struts, and other items used to support a roof or to bear the weight of another part of the structure.

trust. N. A tax entity created by a trust agreement. This entity distributes all or part of its income to beneficiaries as instructed by the trust agreement. This entity is required to pay taxes on undistributed income.

trust account. N. Special account that is used to safeguard the funds of a buyer or seller.

trust deed. N. A conveyance of real estate to a third party to be held for the benefit of another. Commonly used in some states in place of mortgages that conditionally convey title to the lender.

trustee. N. A fiduciary who holds or controls property for the benefit of another.

trustee in bankruptcy. N. Person selected by a judge or creditors of a bankrupt individual to handle matters including the sale of the bankrupt person's assets, management of the funds from the sale of those assets, payment of expenses, and distribution of the balance to creditors. The trustee is usually compensated with a specified percentage of the liquidation sale.

trustee's deed. N. Deed given by a trustee at a deed of trust foreclosure sale.

trustee's sale. N. A foreclosure sale conduced by a trustee under the stipulations of a deed of trust.

trustor. N. Individual creating a trust.

Truth in Lending Act. N. A federal law that requires lenders to fully disclose, in writing, the terms and conditions of a mortgage. A lender should provide the borrower with a Truth in Lending statement within three business days of a loan application. ABBRV. *TILA*.

Truth in Lending statement. N. Document that outlines the costs of the loan, and is given to the borrower so he or she can compare the costs with those of other lenders. The costs listed include the annual percentage rate (APR), finance charge, the amount financed, the payment amount, and the total payments required. The lender is required to give the borrower the final version of his or her statement at or prior to the closing meeting because it is possible that the APR calculated at the time of the loan application could change by closing.

tuckpoint. V. To remove old mortar from between bricks and replace it with new mortar.

Tudor. N. An English-style, imposing-looking house with fortress lines. Siding is chiefly stone and brick with some stucco and half timbers. Windows and doors have molded cement or stone trim around them.

Tudor

turnaround property. N. A deteriorated property that can be restored and sold for a gain.

turnkey project. N. A development in which a developer completes the entire project on behalf of a buyer. The developer turns over the keys to the buyer at completion. All the new tenant or owner has to do is turn the key to the apartment or office in a newly constructed building because everything is completed and ready for occupancy.

turnover. N. (1) Movement of people, such as tenants, in, through, or out of a place. (2) The sale of one average real estate inventory item within a specified time.

two-step mortgage. N. Adjustable mortgage with one interest rate for the first five or seven years of the loan and another for the remainder of the loan term.

two-to-four-family property. N. A piece of property that is represented by one deed, but provides housing for up to four households.

type of use. N. Stipulation in a contract on how a property can or cannot be used.

UCC. ABBRV. Uniform Commercial Code.

UL. ABBRV. Underwriters Laboratories.

ULI. ABBRV. Urban Land Institute.

unadjusted basis. N. The original cost or other basis without regard to salvage value; the basis of property for purposes of figuring depreciation under the accelerated cost recovery system or the modified accelerated cost recovery system.

unaudited opinion. N. An opinion by a certified public accountant who has not audited the relevant financial statements.

underground service. N. Electrical power lines that are run underground in the street and to the building. Underground service has largely replaced the use of aboveground power lines in new housing developments.

underimprovements. N. Property improvements that are below the usual standard expectations.

underlayment. N. A layer of wood between the subfloor and the floor.

underlying mortgage. N. The initial mortgage on a property when other mortgages exist on the same property.

undersupply. N. When demand exceeds the amount of real estate property available, which causes the prices to rise.

Underwriters Laboratories. N. Independent testing group that assesses the safety and quality of various electrical components. Once a product is approved, the manufacturer can place a "UL" label on the item. ABBRV. *UL*.

underwriting. N. Process by which a lender determines the risk of a loan, which involves an analysis of the borrower's creditworthiness.

underwriting fee. N. Fee charged by the lender to do the work to verify information necessary to make a decision as to whether or not to approve a loan; typically included in the borrower's application fee.

undisclosed heir. N. One who claims, after the death of an owner lacking a will, a right to a piece of the owner's property.

undisclosed principal. N. (1) A major party to a transaction who remains anonymous. (2) Principal that has not been disclosed to one party in a transaction.

undisclosed spouse. N. A marital partner who is not mentioned in a will, but who can claim a right to a piece of property.

undivided interest. N. Ownership by two or more persons, which entitles each to the right to use the entire property.

undue influence. N. When a person is forced to perform in a certain manner due to excessive influence, pressure, or fear brought by another party; may be used to void a contract.

unearned income. N. Taxable income other than that received for services performed (i.e., earned income). Unearned income includes money received for the investment of money or other property, such as interest, dividends, and royalties. It also includes pensions, alimony, unemployment compensation, and other income that is not earned.

unearned increment. N. An increase in the value of a property, which is not due to any effort on the part of the owner.

unencumbered property

unencumbered property. N. Real estate that is free and clear of liens and obligations, such as a house without a mortgage.

unenforceable contract. N. Agreement that cannot be legally enforced by the courts, such as contracts with minors and those containing fraud.

Uniform Commercial Code. N. Group of laws that standardize the state laws applicable to commercial transactions. ABBRV. *UCC*.

Uniform Residential Appraisal Report. N. The standard form used for reporting the appraisal of a dwelling, which provides numerous checklists, definitions, and certifications. ABBRV. *URAR*.

Uniform Residential Landlord and Tenant Act. N. Legislation dealing with landlord-tenant contracts and relationships; provides protection to renters in thirteen states.

Uniform Settlement Statement. N. Form prescribed by the Real Estate Settlement Procedures Act for federally related mortgages, which must be prepared at the closing of title and contain certain relevant closing information. Both the buyer and seller are given copies.

Uniform Standards of Professional Appraisal Practice. N. Set of requirements formulated and adhered to by the members of the Appraisal Foundation. ABBRV. *UNPAP*.

unilateral contract. N. One-sided contract where, even if one party makes a promise, the second party is not legally required to perform, but may do so, thus obligating the first party.

unilateral listing. N. Listing of property that is open and has no one real estate agent who has the sole right to sell the property.

unimproved land. N. Raw land in its natural state that has no installed improvements or structures.

unincorporated group. N. Association of persons not treated as a corporation, such as a limited partnership.

unit. N. Segregated part of a structure, i.e., one apartment in an apartment building or an office in a commercial building.

unit cost. N. The cost of a single item on a construction budget; for example, one 2 x 4 x 8' stud. The unit cost is multiplied by the number of units required for the job to determine the total cost.

unities. N. The four characteristics of interest, possession, time, and title, which must unite in order to form a joint tenancy.

unit in place method. N. Valuation of property based on its replacement cost; cost of major elements per square foot multiplied by amount of square feet is the estimate of building cost.

unity. N. Legal dictate necessary for property to be owned as joint tenants.

unity of interest. N. Principle that joint tenants must acquire their interest by the same conveyance and with the same interest.

unity of possession. N. Principle that joint tenants must have equal rights to possession of the entire property.

unity of time. N. Principle that joint tenants must acquire their title simultaneously.

unity of title. N. Principle that joint tenants must acquire their interest from the same deed or will.

unlawful detainer. N. Illegally holding on to or keeping the property of another.

unleveraged. N. Operating without the use of borrowed money.

unlike properties. N. Properties that are used for different purposes or are of different types.

unqualified audit. N. A complete audit.

unqualified opinion. N. Auditor's opinion of a financial statement, given without any reservations.

unrealized gain. N. Increase in the value of property while it is being held. Gain is only realized upon sale.

unrecorded deed. N. Deed transferring ownership from one person to another, but is not officially recorded.

unsecured loan. N. A loan that is not backed by collateral.

unsecured note. N. Credit note with only the borrower's financial situation and credit history as security.

unsolicited listing. N. Real estate listing obtained without any effort on the part of the real estate broker.

unstated interest. See *imputed interest*.

up-front fee. N. Fee received immediately when a contract is signed or an investment is made.

upgrades. N. Options offered to buyers that go beyond the standard. In a new home, they are often more lighting, better carpeting, etc.

upset date. N. Contract provision allowing a purchaser the right to cancel a contract if the occupancy requirements are not satisfied by a certain date.

upset price. N. An established amount in a bidding procedure or auction, below which the seller is not obligated to accept the winning bid. Synonymous with *reserve price*.

upside potential. N. An approximation of the potential appreciation of value in real estate by considering location, amenities, or increase in rental income.

upzoning. N. Rezoning of property to a higher use.

URAR. ABBRV. Uniform Residential Appraisal Report.

urban development action grant. N. Grant intended to stimulate private investment in distressed cities and urban areas by providing federal seed money, thereby attracting private funds for revitalization. U.S. Department of Housing and Urban Development funds are transferred directly to cities and states for urban renewal projects.

Urban Land Institute. N. Nonprofit entity from which persons can obtain data and advice on the best utilization of land. ABBRV. *ULI.*

urban property. N. Real estate located in a heavily populated city area.

urban renewal. N. Older property purchased by a governmental agency and improved or modernized, or demolished and replaced.

urban space. N. Land availability in an urban area.

urban sprawl. N. Unplanned and unexpected expansion of a large area of development in an urban area.

U.S. Department of Housing and Urban Development. N. Federal agency overseeing the Federal Housing Administration, in addition to a number of housing and community development programs. ABBRV. *HUD.*

use. N. A term in a lease that describes and limits the tenant's uses of the space in question.

useful life. N. Usual operating service life of property used for depreciation accounting, which does not necessarily coincide with the actual physical life.

use tax. N. Levy charged for the use of things such as town water, etc.

USPAP. ABBRV. Uniform Standards of Professional Appraisal Practice.

U.S. Tax Court. N. One of three trial courts of original jurisdiction that decide litigation involving federal income, death, and gift taxes.

usufructuary right. N. One individual's right to utilize the property of another, such as the privileges given under an easement.

usury. N. Excessive and illegal interest charged on a loan.

usury laws. N. State laws that limit the interest rate charged to persons borrowing money in that state.

utilities. N. Services provided to land or buildings by public utility companies, such as water, gas, electricity, etc.

utilities availability. N. Promise by a landlord in a lease guaranteeing necessary utilities to the tenant.

utility easement. N. A passage through property that is granted by the owner to a public utility.

utility room. N. Room that contains the appliances necessary for the maintenance of that establishment.

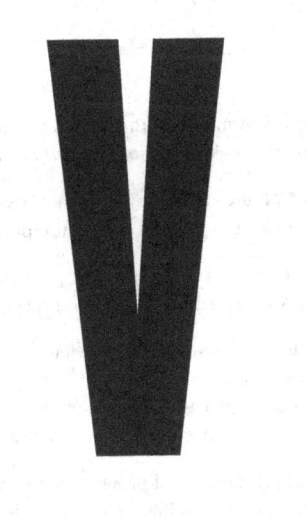

VA. ABBRV. U.S. Department of Veterans Affairs.

VA guarantee. N. Loan made by private lenders that allows qualified veterans to buy a house with no down payment based on certain conditions; the Department of Veterans Affairs (VA) can guarantee up to $417,000 of the total loan. The qualification guidelines for VA loans are more flexible than those for either the Federal Housing Administration or conventional loans.

VA loan. N. A loan made through the Department of Veterans Affairs.

VA mortgage. N. A mortgage that is guaranteed by the Department of Veterans Affairs. See also *government mortgage*.

vacancy and credit losses. N. Losses incurred by the owner due to the tenant's failure to pay rent or because property is unoccupied.

vacancy factor. N. Estimated percentage of rental that will not be made because of actual and anticipated vacancies.

vacancy rate. N. Percentage of unoccupied rental property. Idle space causes significant cash drain when cash inflows are not received to offset the cash outflows of maintenance. Most properties have a minimum occupancy rate to break even.

vacant land. N. Unoccupied property not currently being used. It may have utilities and off-site improvements, as contrasted with raw land that has no improvements or structures.

vacate. V. To leave a property. If the legally agreed-upon term is not fulfilled, vacating a property does not relieve the occupant of liability.

vacation home. N. A home owned in addition to a person's primary residence. Interest and real estate taxes are tax deductible with the Internal Revenue Service, though certain conditions apply.

valid. ADJ. Factual; representations made by a realtor to a prospective purchaser fall into this category.

valid contract. N. An agreement that is legally binding because it is in conformity with all legal requirements and conditions.

valid deed. N. Legally proper instrument that transfers title of real estate from seller to buyer.

valuable consideration. N. A promised payment that can have a claim enforced upon it if the promisor is unwilling to act on it.

valuation. N. Estimated worth of a property as valued through an appraisal.

value. N. An expression of monetary worth of a particular piece of real estate; what a buyer will pay for a property. Synonymous with *price*.

value after the taking. N. The value left on a property after it has been condemned.

value before the taking. N. Market price of a property prior to a condemnation proceeding.

value in exchange. N. The giving of money, goods, or services in exchange for other goods or services provided by another party.

value in use. N. Discounted value of net cash to be obtained from a property calculated by consideration of annual cash inflows plus the disposal value.

variable interest rate. N. Loan rate that changes based on fluctuations in the rate paid on Treasury bills or bank certificates of deposit.

variable maturity mortgage. N. A long-term mortgage loan with an interest rate that may be adjusted periodically by the lender.

Payment levels may remain the same, but the loan maturity is lengthened or shortened to achieve the adjustment.

variable rate. N. Interest rate that changes periodically in relation to an index.

variable rate mortgage. N. A loan with an interest rate that adjusts with the changes in rates paid on Treasury bills or bank certificates of deposit. ABBRV. *VRM*. See also *adjustable rate mortgage*.

variance. N. An allowable deviation from the land use prescribed by the existing zoning ordinances.

vaulted ceiling. N. An arched and sometimes beamed ceiling.

vendee. N. A buyer of real property.

vendee's lien. N. Legal right of a purchaser of a piece of real estate to the paid purchase price plus direct costs of acquisition if the seller fails to render the deed to the property.

vendor. N. A seller of real estate or other products.

vendor's lien. See *purchase money mortgage*.

venture capital. N. Financing source for new businesses or turn-around ventures with high risk/high reward possibilities. Venture capital can be anything from seed money to full financing; sources include wealthy individuals, limited partnerships, and business investment companies.

venue. N. Geographic location of a court action, which should occur in the place where jurisdiction applies. Change of venue, which is the move to another locality, occurs in a criminal action when a fair trial is precluded.

verification. N. Sworn statement made before a qualified officer that the contents of a document are correct.

Victorian style

verification of deposit. N. A document signed by the borrower's bank or other financial institution verifying the borrower's account balance and history.

verification of employment. N. Statement by a borrower's employer that confirms the borrower's salary and position and is part of the loan process. ABBRV. *VOE*.

verify. V. To collect supporting evidence; to prove.

vested. N. Having the right to use a portion of a fund such as an individual retirement fund. For example, individuals who are 100% vested can withdraw all of the funds that are set aside for them in a retirement fund. However, taxes may be due on any funds that are actually withdrawn.

vestibule. N. A small entrance hall or room.

Veterans Administration. N. A program of the Department of Veteran Affairs that allows most veterans to purchase a house without a down payment. ABBRV. *VA*.

vicarious liability. N. The responsibility of one person for the actions of another.

Victorian style. N. Architectural style of the mid-nineteenth century.

Victorian style

vinyl-clad windows. N. Wood windows sheathed in vinyl on the outside.

violation. N. An act or condition contrary to the permitted use of a property.

visual right. N. Occupant's right to see out of a window without being hindered.

voidable contract. N. A contract that may be rejected by either of the parties to it.

volatile market. N. Sudden and unpredictable short-term price movements in real estate.

voluntary alienation. N. The conveyance of title between two parties or businesses by use of a deed.

voluntary bankruptcy. N. The declaration of bankruptcy by an insolvent person or business.

voluntary conveyance. See *deed-in-lieu*.

voluntary lien. N. Lien that a homeowner willingly gives to a lender, such as a mortgage.

VRM. ABBRV. Variable rate mortgage.

wainscotting

wainscotting. N. Wood paneling, tongue-and-groove boards, or similar material installed between a baseboard and a chair rail.

waiver. N. Voluntary surrender of some right or privilege, usually in writing.

waiver of lien. N. Voluntarily relinquishing the right of a lien, usually temporarily. This waiver may be explicitly stated or implied.

waiver of subrogation. N. Waiver that protects a first and second party from being sued for the recovery of damages by third parties.

waiver of tax lien. N. A form signed by a taxing jurisdiction certifying it will not file a lien against a property owner for back taxes. This form comes into play when real estate is being sold by an estate after the owner's death.

walkaway risk. N. Risk that occurs when a buyer or seller decides not to go through with a transaction. If property is not sold at the offered price, that price may not be realized at a later date, a risk that applies to both the buyer and seller.

walk-out basement. N. Type of basement that allows a door to open into the yard because the basement is at ground level.

walk-through inspection. N. Final inspection of a home to check for problems that may need to be corrected before closing, usually done the day before.

walkup. N. Building of two or more floors with no elevator. This type of building is usually found in cities.

warehouse fee. N. The lender's cost of holding a borrower's loan temporarily before it is sold on the secondary market. It usually involves a closing cost fee.

warehouse property. N. Commercial structure used to hold products and goods for a fee; typically located in an industrial area.

warehousing. N. (1) Loans made by banks to other lenders for their underwritten stocks or bonds. These stocks and bonds are issued to both household and institutional investors for their portfolios. (2) When lenders store up loans before packaging them for sale in the secondary market.

warranted price. N. Price that is considered a fair amount for a real estate transaction, justified by the conditions involved in the exchange.

warranty. N. (1) A statement of fact or opinion about a company's financial condition. (2) An agreement between a buyer and a seller of goods or services, detailing the conditions under which the seller will make repairs or fix problems without cost to the buyer. Many homes—both old and new—carry warranties as a sales incentive.

warranty deed. N. Deed that assures that the title to a property is free of any legal claims or encumbrances and includes covenants of seizing, express warranties of title, right to quiet enjoyment, and freedom from encumbrances.

warranty insurance. N. Insurance on a property covering repair and/or replacement costs in the event of damage or loss. Warranties vary considerably as to what is covered and the duration of the policy.

warranty of habitability. N. Implied assurance from the landlord to the tenant that an apartment is safe and free from any hazard.

wasting asset. N. A fixed asset that has a limited useful life, thus declines in value over time. Natural resources are typically considered wasting assets because they are extracted at a faster pace than nature can replace them.

water rights. See *riparian rights*.

waterfront property. N. Structure adjacent to a lake or other type of water, which has a higher sale value due to greater demand.

watershed. N. (1) Land area where water collects. (2) A dividing point that sends water runoff flowing into different drainage areas.

weak market. See *depressed market*.

wear and tear. N. Decline in the value of a property due to physical damage, old age, or environmental factors.

weatherhead. See *entrance cap*.

well. N. (1) A deep hole or shaft sunk into the earth to tap and remove an underground supply of water or natural gas. (2) A shaft in a building or between buildings that is open to allow access for light and air.

wellhold. N. The space that holds a staircase and the space around it.

western framing. N. House framing that has the wall framing resting on top of the subfloor with each story being built up as a separate unit. Balloon framing, another type of western framing, has studs running from the bottom of the first floor to the top of the second floor. Synonymous with *platform framing*.

Western Row House. N. A nineteenth-century-style house built to cover an entire street or block. Synonymous with *Western Town House*.

Western Town House. See *Western Row House*.

wetlands. N. Land situated near water that meets various criteria regarding water level, soil, and plant growth. Wetlands are subject to extreme levels of government regulation regarding building activity. They may include swamps, marshes, and floodplains.

what-if analysis. N. An affordability analysis based on predictions. A what-if scenario can include changes to monthly income, debts, or down payment funds, or to the qualifying ratios or down payment expenses that are used in the analysis. A what-if scenario can be used to explore different ways to improve a borrower's ability to afford a house. Many lending and personal finance websites offer ways to run this analysis.

white elephant. N. A property considered undesirable in the marketplace, i.e., too expensive to maintain or unable to generate enough rent to pay for itself.

wild deed. N. An improperly recorded deed.

will. N. Legal document that outlines the disposition of a deceased person's estate. Synonymous with *testament*.

Williamsburg Georgian. N. English-style house representative of the early colonial houses built in America during the early 1700s. They had simple exterior lines and fewer of the decorative devices seen in the later Georgian style houses. Williamsburg designs were two- and three-story rectangular houses with two large chimneys rising high above the roof at either end of the house.

window light. N. An individual pane of glass.

Window light

window seat. N. A bench built under an interior window.

window well. N. A curved, corrugated steel insert used to isolate basement windows from moisture if they are under the soil line.

window well covers. N. Curved plastic covers formed to install over an open window well.

winterize. V. To adapt property or machinery for operation or storage during the cold weather months.

without recourse. N. Phrase used to endorse a note or bill that indicates that the holder cannot look to the debtor personally for compensation if the loan is not paid. In that case, the holder's only recourse is to go after debtor's property.

witness. N. An individual who provides evidence under penalty of perjury, under oath in a trial. V. To observe an event, transaction, or contract.

workers' compensation insurance. N. Government-mandated insurance that covers job-related injuries to employees and is paid for by employers.

working capital. N. Current assets minus current liabilities. Synonymous with *net current assets*.

working drawing. N. A drawing made to scale with details and markings to facilitate use by builders and engineers during construction.

work letter. N. A document that spells out improvements that a landlord must make before a tenant will accept the property, usually in the context of a commercial property. This letter will sometimes include drawings, pricing, and specific approval dates as exhibits.

workout. N. Situation in which a lender loosens some terms for overdue payment when a borrower runs into trouble paying back a loan. The lender may lower an interest rate or lengthen the time of repayment so the borrower can work out the problems paying the debt.

worksheet. N. A record of compiled information that is generally not sent to the Internal Revenue Service with a tax return.

workshop. N. Area used for working with materials or machinery.

wraparound mortgage. N. A mortgage that includes the remaining balance on an existing first mortgage plus an additional amount requested by the mortgagor. Full payments on both mortgages are made to the wraparound mortgagee, who then forwards the payments on the first mortgage to the first mortgagee.

writ. N. Court order requiring a person or business to react in a specific manner.

write down. N. A downward adjustment made by an owner in the accounting value of an asset.

write-off. V. To charge an asset amount to expense or loss in order to reduce the value of that asset and one's earnings.

writ of ejectment. N. Court order allowing a landlord to evict a tenant because of nonpayment of rent or damaging property. The writ contains needed instructions and directs an officer of the court to execute it.

writ of execution. N. Court order allowing the seizure and sale of property due to nonpayment of taxes or foreclosure of property.

writ of mandamus. N. Court order that stops or directs a judicial directive.

written-down value. See *depreciated cost*.

yard. N. Open ground area on the property.

year-over-year. ADJ. Compared to the same time period in the previous year.

year to date. N. Time period that starts January 1 of the current year and ends today. ABBRV. *YTD*.

yield. N. (1) The return on an investment; volume or amount produced. (2) A permanent deformation in a material caused by its being bent or stretched.

yield curve. N. A graph that shows current interest rates of similar obligations ranked by maturity.

yield to maturity. N. The average rate of return on a bond with a maturity of at least one year if it is held to its maturity date and if all cash flows are reinvested at the same rate of interest. It includes an adjustment for any premium paid or discount received. Yield to maturity is used to compare the relative values of different types of investments, including bonds and real estate. ABBRV. *YTM*.

YTD. ABBRV. Year to date.

YTM. ABBRV. Yield to maturity.

zero lot line. N. Lot where a home is set to the lot boundary, which leaves very little space between the houses.

zero net. N. Occurs when there is no money left for the seller from the sale of a property after all outstanding expenses are paid.

zero rate loan. N. Loan with a large down payment and a balance paid in equal periodic payments over a short period of time, with no interest charged. Usually offered by an eager seller.

zone. V. To set off an area of land by ordinance for a specific purpose. Changing that purpose typically requires application to a zoning board of appeals.

zoning. N. Rules and regulations that control the use of land, which is broken down into districts, and determine how private property is to be used or what construction is allowed. Zoning may be either commercial, residential, industrial, or agricultural. Zoning also restricts height limitations, noise, parking, and open space. Residential zoning may consist of single-family dwellings, two-family dwellings, or apartments.

zoning board of appeals. N. Local government board that is used to resolve zoning disputes.

zoning laws. N. Ordinances created by local government to cover real estate development, including structural and asthetic points. They usually define usage classifications from agricultural to industry, and also building restrictions such as minimum and maximum square footage requirements, violations penalties, and procedures.

zoning map. N. Map that shows locality divided into districts, shows status and usage of each district, and is kept current.

zoning ordinances. N. Regulations determined by each municipality to establish different zoning restrictions and classifications;

for example, building height, building type, etc. Penalties are assessed for violations of zoning ordinances.

zoning records. N. Compilation of zoning requirements and any changes made to them.

zoning variance. N. A modification of the existing zoning law, which can be made in certain instances; for example, to build a taller building or a different type of housing than that which is allowed.

ABBREVIATIONS

Abbreviations

A/C. Air conditioner or air conditioning.

ADA. Americans with Disabilities Act.

ACRS. Accelerated Cost Recovery System.

ADR. Asset depreciation range.

AGI. Adjusted gross income.

AI. Appraisal Institute.

AIA. American Institute of Architects.

ALTA. American Land Title Association.

APA. American Planning Association.

APR. Annual percentage rate.

APY. Annual percentage yield.

ARM. Adjustable rate mortgage.

ASA. American Society of Appraisers.

ASHI. American Society of Home Inspectors.

ASREC. American Society of Real Estate Counselors.

ASTM. American Society for Testing Materials.

BPI. Buying Power Index.

CADD. Computer Aided Design and Drafting.

CAE. Certified Assessment Evaluator.

CAGR. Compound Annual Growth Rate.

CD. Certificate of deposit.

CERCLA. Comprehensive Environmental Response, Compensation, and Liability Act.

CMA. Comparative market analysis.

CO. Certificate of Occupancy.

COFI. Cost of Funds Index.

CPI. Consumer Price Index.

DBA. Doing business as.

DRM. Direct reduction mortgage.

DSCR. Debt service coverage ratio.

EBIT. Earnings before interest and taxes.

EBITDA. Earnings before interest, taxes, depreciation and amortization.

ECOA. Equal Credit Opportunity Act.

ELR. Equivalent level rate.

EPA. Environmental Protection Agency.

et al. And others.

et con. With husband.

et ux. With wife.

Fannie Mae. Federal National Mortgage Association.

FAR. Floor area ratio.

FASB. Financial Accounting Standards Board.

FDIC. Federal Deposit Insurance Corporation.

FHA. Federal Housing Administration.

FHLBB. Federal Home Loan Bank Board.

FNMA. Federal National Mortgage Association.

Freddie Mac. Federal Home Loan Mortgage Corporation.

FSBO. For sale by owner.

GAAP. Generally accepted accounting principles.

Abbreviations

GDP. Gross domestic product.

GEM. Growing equity mortgage.

Ginnie Mae. Government National Mortgage Association.

GLA. Gross leasable area.

GNMA. Government National Mortgage Association.

GPM. Graduated payment mortgage.

GRI. Graduate Realtor Institute.

HECM. Home equity conversion mortgage.

HELOC. Home equity line of credit.

HOW. Homeowners' warranty.

HUD. U.S. Department of Housing and Urban Development.

HVAC. Heating, ventilation, and air conditioning.

IRC. Internal revenue code.

IRR. Internal rate of return.

IRS. Internal Revenue Service.

LIBOR. London InterBank Offered Rate.

LTV. Loan-to-value ratio.

MACRS. Modified accelerated cost recovery system.

MGIC. Mortgage Guaranty Insurance Corporation.

MSA. Metropolitan statistical area.

NAA. National Apartment Association.

NAEBA. National Association of Exclusive Buyers Agents.

NAHB. National Association of Home Builders.

NAHI. National Association of Home Inspectors.

NAIFA. National Association of Independent Fee Appraisers.

NAR. National Association of Realtors®.

NAREIT. National Association of Real Estate Investment Trusts.

NCREIF. National Council of Real Estate Investment Fiduciaries.

NFIP. National Flood Insurance Program.

NIMBY Not in my backyard.

NLA. Net leasable area.

NNN. Triple net lease.

NOI. Net operating income.

NOL. Net operating loss.

NPV. Net present value.

NSREA. National Society of Real Estate Appraisers.

OILSR. Office of Interstate Land Sales Registration.

OSHA. Occupational Safety and Health Administration.

OTS. Office of Thrift Supervision.

PAM. Pledged account mortgage.

PITI. Principal, interest, taxes, and insurance.

PLAM. Price level adjusted mortgage.

PMI. Private mortgage insurance.

PUD. Planned unit development.

REIT. Real estate investment trust.

REMIC. Real Estate Mortgage Investment Conduit.

REO. Real estate owned.

RESPA. Real Estate Settlement Procedures Act.

Abbreviations

RFP. Request for proposal.

ROA. Return on assets.

ROE. Return on equity.

ROI. Return on investment.

ROIC. Return on invested capital.

SAM. Shared appreciation mortgage.

SBA. Small Business Administration.

SIOR. Society of Industrial and Office Realtors.

SREA. Society of Real Estate Appraisers.

SRS. Specialist in Real Estate Securities.

SYD. Sum-of-Years'-Digits.

TILA. Truth in Lending Act.

UCC. Uniform Commercial Code.

UL. Underwriters Laboratories.

ULI. Urban Land Institute.

URAR. Uniform Residential Appraisal Report.

USPAP. Uniform Standards of Professional Appraisal Practice.

VA. Veterans Administration.

VRM. Variable rate mortgage.

WEBSITES

Websites

American Institute of Architects:
www.aia.org

American Land Title Association
www.alta.org

American Motel Hotel Brokers Network
www.amhbnetwork.com

American Planning Association
www.planning.org

American Society of Appraisers
www.appraisers.org

American Society of Farm Managers and Rural Appraisers
www.asfmra.org

American Society of Home Inspectors
www.ashi.com

American Society for Testing Materials
www.astm.org

Appraisal Institute
www.appraisalinstitute.org

Counselors of Real Estate
www.cre.org

Equifax
www.equifax.com

Experian
www.experian.com

Federal Trade Commission
www.ftc.gov

Internal Revenue Service
www.irs.gov

National Apartment Association
www.naahq.org.

National Association of Exclusive Buyer Agents
www.naeba.org

National Association of Home Builders
www.nahb.org

National Association of Home Inspectors, Inc.
www.nahi.org

National Association of Independent Fee Appraisers
www.naifa.com

National Association of Real Estate Investment Trusts
www.reit.com

National Association of Realtors®
www.realtor.org

National Association of Review Appraisers and Mortgage Underwriters
http://naramu.org

National Council of Real Estate Investment Fiduciaries
www.ncreif.com

National Society of Real Estate Appraisers
www.nsrea.org

Real Estate Index
www.realestateindex.com

Small Business Administration
www.sba.gov

Society of Industrial and Office Realtors
www.sior.com

TransUnion
www.transunion.com

Urban Land Institute
www.uli.org

U.S. Department of Housing and Urban Development
www.hud.gov

Veterans Administration
www.homeloans.va.gov